Mathematics
as Problem Solving

Second Edition

Alexander Soifer

Mathematics as Problem Solving

Second Edition

Springer

Alexander Soifer
College of Letters, Arts and Sciences
University of Colorado at Colorado Springs
1420 Austin Bluffs Parkway
Colorado Springs, CO 80918
USA
asoifer@uccs.edu

ISBN 978-0-387-74646-3   ISBN: 978-0-387-74647-0 (eBook)
DOI 10.1007/978-0-387-74647-0

Library of Congress Control Number: 2009921736

Mathematics Subject Classification (2000): 00-XX, 00A05, 00A07, 00A08, 00A35, 97A20, 05CXX, 05C15, 05C55, 05-XX

Cover designed by Mary Burgess

Printed on acid-free paper

springer.com

To Mark and Julia Soifer

Frontispiece reproduces the front cover of the original edition. It was designed by my later father Yuri Soifer, who was a great artist. Will Robinson, who produced a documentary about him for the Colorado Springs affiliate of ABC, called him "an artist of the heart." For his first American one-man show at the University of Colorado in June–July 1981, Yuri sketched his autobiography:

I was born in 1907 in the little village Strizhevka in the Ukraine. From the age of three, I was taught at the Cheder (elementary school by a synagogue), and since that time I have been painting. At the age of ten, I entered Feinstein's Jewish High School in the city of Vinniza. The art teacher, Abram Markovich Cherkassky, a graduate of the Academy of Fine Arts at St. Petersburg, looked at my book of sketches of praying Jews, and consequently taught me for six years, until his departure for Kiev. Cherkassky was my first and most important teacher. He not only critiqued my work and explained various techniques, but used to sit down in my place and correct mistakes in my work until it was nearly unrecognizable. I couldn't then touch my work and continue – this was unforgettable.

In 1924, when I was 17, my relative, the American biologist, who later won the Nobel Prize in 1952, Selman A. Waksman, offered to take me to the United States to study and become an artist, and to introduce me to Chagall, but my mother did not allow this, and I went to Odessa to study at the Odessa Institute for the Fine Arts in the studio of Professor Mueller. Upon graduation in 1930, I worked at the Odessa State Jewish Theater, and a year later became the chief set and costume designer. In 1934, I came to Moscow to design plays for Birobidzhan Jewish Theater under the supervision of the great Michoels. I worked for the Jewish newspaper **Der Emes,** *the Moscow Film Studio, Theater of Lenin's Komsomol, and a permanent National Agricultural Exhibition. Upon finishing my 1941–1945 service in World War II, I worked for the National Exhibition in Moscow, VDNH.*

All my life, I have always worked in painting and graphics. Besides portraits and landscapes in oil, watercolor, gouache, and marker (and also acrylic upon the arrival in the USA), I was always inspired (perhaps, obsessed) by the images and ideas of the Russian Civil War, Word War II, biblical stories, and the little Jewish village that I came from.

The rest of my biography is in my works!

Front cover of the first edition, 1987, by Yuri Soifer.

Foreword

This book joins several other books available for the preparation of young scholars for a future that involves solving mathematical problems.

This training not only increases their fitness in competitions, but may also help them in other endeavors they may engage in the future.

The book is a diversified collection of problems from all areas of high school mathematics, and is written in a lively and engaging way.

The introductory explanations and worked problems help guide the reader without turning the additional problems into rote repetitions of the solved ones.

The book should become an essential tool in the armamentarium of faculty involved with training future competitors.

Branko Grünbaum
Professor of Mathematics
University of Washington
June 2008, Seattle, Washington

Foreword

This was the first of Alexander Soifer's books, I think, preceding *How Does One Cut a Triangle?* by a few years. It is short on anecdote and reminiscence, but there is charm in its youthful brusqueness and let's-get-right-to-business muscularity. And, mainly, there is a huge lode of problems, very good ones worked out and very good ones left to the reader to work out.

Every mathematician has his or her bag of tricks, and perhaps every mathematician will find some part of this book to view with smug condescension, but there may not be a mathematician alive that can so view all of this book. I notice that Paul Erdős registered his admiration for the chapters on combinatorics and geometry. For me, the Pigeonhole Principle problems were fascinating, exotic, and hard, and I would like to base a course on that section and on parts of the chapters on combinatorics and geometry.

Anyone coaching a Putnam Exam team should have a copy of this book, and anyone trying out for a Putnam Exam team would do well to train with this book. Training for prize exams is a good entree to higher mathematics, but even if you are not a competitive type, this book could well be the portal that will lead you into the wonderful world of mathematics.

Peter D. Johnson, Jr.
Professor of Mathematics
Auburn University
June 12, 2008, Auburn, Alabama

Foreword

In *Mathematics as Problem Solving*, Alexander Soifer has given an approach to problem solving that emphasizes basic techniques and thought rather than formulas. As he writes in the introduction to Chapter 2 (Numbers),

> *Numerous beautiful results could be presented here, but I will limit myself to problems illustrating some ideas and requiring practically no knowledge of number theory.*

The chapter headings are

- Language and a Few Celebrated Ideas
- Numbers
- Algebra
- Geometry
- Combinatorial Problems

Each topic is suitable for high school students, and there is a pleasant leanness to the list of topics (compare this with a current calculus text). The Chinese Remainder Theorem is out; the Pigeonhole Principle is in. As the reader will at some point discover, the Chinese Remainder Theorem can be deduced from the Pigeonhole Principle. Now is the time for fundamental problem solving; first things first. At the same time, nontrivial ruler and compass construction problems are basic to a proper understanding of geometry. Dr. Soifer has made a wise choice to emphasize this topic.

The 200 or so problems are well chosen to go with the emphasis on fundamental techniques, and they provide a rich resource. Some of the problems are appropriately routine, while some others are "little results" found by mathematicians in the course of their research. For example, Problem 1.29 is a rewording of a result mentioned in a survey paper by Paul Erdős; the discovery was originally made by Erdős and V.T. Sós. This problem also appeared on the 1979 USA Mathematical Olympiad.

> **1.29** (First Annual Southampton Mathematical Olympiad, 1986) An organization consisting of n members ($n > 5$) has $n + 1$ three-member committees, no two of which have identical membership. Prove that there are two committees in which exactly one member is common.

Mathematics as Problem Solving is an ideal book with which to begin the study of problem solving. After readers have gone on to study more comprehensive sources, *Mathematics as Problem Solving* is likely to remain in a place of honor on their bookshelf.

Cecil Rousseau
Professor of Mathematics
Memphis State University
June 2008, Memphis, Tennessee

Preface to the Second Edition

> *The moving power of mathematical invention is not reasoning but imagination.*
>
> Augustus de Morgan

I released this book over twenty years ago. Since then she lived her own life, quite separately from me. Let me briefly trace her life here.

In March 1989, her title, *Mathematics as Problem Solving*, became the first "standard for school mathematics" of the National Council of Teachers of Mathematics [2]. In 1995, her French 4000-copy edition, *Les mathématiques par la résolution de problèmes, Éditions du Choix*, quickly sold out.

She was found charming and worthy by Paul Erdős, Martin Gardner, George Berszenyi, and others:

> The problems faithfully reflect the world-famous Russian school of mathematics, whose folklore is carefully interwoven with more traditional topics. Many of the problems are drawn from the author's rich repertoire of personal experiences, dating back to his younger days as an outstanding competitor in his native Russia and spanning decades and continents as an organizer of competitions at the highest level. – George Berzsenyi

> The book contains a very nice collection of problems of various difficulties. I particularly liked the problems on combinatorics and geometry. – Paul Erdős

> Professor Soifer has put together a splendid collection of elementary problems designed to lead students into significant mathematical concepts and techniques. Highly recommended. – Martin Gardner

In the "extended" *American Mathematical Monthly* review, Cecil Rousseau paid her a high compliment:

> *Retelling the best solutions and sharing the secrets of discovery are part of the process of teaching problem solving. Ideally, this process is characterized by mathematical skill, good taste, and wit. It is a characteristically personal process and the best such teachers have surely left their personal marks on students and readers. Alexander Soifer is a teacher of problem solving and his book,* Mathematics as Problem Solving, *is designed to introduce problem solving to the next generation.*

This poses a problem: how does one reach out to the next generation and charm it into reading and doing mathematics? I am deeply grateful to Ann Kostant for solving this problem by inviting a new edition of this book into the historic Springer. I thank Col. Dr. Robert Ewell for converting my sketches into real illustrations. I am so very grateful to the first readers of this manuscript, Branko Grünbaum, Peter D. Johnson, Jr., and Cecil Rousseau for their comments and forewords.

For the expanded Springer edition, I have added a sixth chapter dedicated to my favorite problem of the many problems that I have created, "Chess 7×7." I found three beautiful solutions to it. Moreover, this problem was inspired by the "serious" mathematics of Ramsey Theory, and once it was solved, it led me back to the "serious" mathematics of finite projective planes. I hope you will enjoy this additional chapter.

Let me mention for those who would like to read my other book that this book was followed by the books [9, 1, 10] listed in the bibliography. Then there came *The Mathematical Coloring Book* [11], after 18 years of writing. Books [12] and [13] will follow soon, as will new expanded editions of the books [9, 1, 10]. All my books will be published by Springer.

Write back to me; your solutions, problems, and ideas are always welcome!

Alexander Soifer
Colorado Springs, Colorado
May 8, 2008

Preface to the First Edition

> *Remember but him, who being demanded, to what*
> *purpose he toiled so much about an Art, which could*
> *by no means come to the knowledge of many. Few are*
> *enough for me; one will suffice, yea, less than one will*
> *content me, answered he. He said true: you and another*
> *are a sufficient theatre one for another; or you to your*
> *selfe alone!!*
>
> Michel de Montaigne
> *Of Solitarinesse*. Essayes [6]

I was fortunate to grow up in the problem-solving atmosphere of Moscow with its mathematical clubs, schools, and Olympiads. The material for this book stems from my participation in numerous mathematical competitions of all levels, from school to national, as a competitor, an organizer, a judge, and a problem writer; but most importantly, from the mathematical folklore I grew up on.

This book contains about 200 problems, over one-third of which are discussed in detail, sometimes even with two or more solutions. When I started, I thought that beauty, challenge, elegance, and surprising results and solutions alone would determine my choices. During my work, however, one more factor powerfully forced itself into account: the interplay of selected problems.

This book is written for high school and college students, teachers, and everyone else desiring to experience the mystery and beauty of mathematics. It can be and has been used as a text for an undergraduate or graduate course or workshop on problem solving.

Auguste Renoir once said that just as some people all their lives read one book (the Bible, for example), so could he paint all his life one painting. I cannot agree with him more. This is the book I am going to write all my life. That is why I welcome so much your comments, corrections, ideas, alternative solutions, and suggestions to include other methods or to cover other areas of mathematics. Do send me your ideas and solutions: best of them as well as the names of their authors will be included in the future revised editions of this book. I hope, though, that this book will never reach the intimidating size of a calculus text.

One can fairly make an argument that this book is raw, unpolished. Perhaps that is not all bad: sketches by Modigliani give me, for one, so much more than sweated-out oils of Old Masters. Maybe a problem-solving book ought to be a sketch book!

To assign true authorship to these problems is as difficult as to folklore tales. The few references that I have given indicate my source rather than a definitive reference to the first mentioning of a problem. Even problems that I created and published myself might have existed before I was born!

I thank Valarie Barnes for bravely agreeing to type this manuscript; it was her first encounter of the mathematical kind. I thank my student Richard Jessop for producing such a masterpiece of typesetting art.

I am grateful to my parents Yuri and Rebbeca for introducing me to the world of arts, and to my children Mark and Julia for inspiration. My special thanks go to the first judges of this manuscript, my students in Colorado Springs and Southampton for their enthusiasm, ideas, and support.

A. Soifer
Colorado Springs, Colorado
November 1986

Contents

1

Language and Some Celebrated Ideas

1.1 Streetcar Stories

I would like to start our discussion with the following stories.

Streetcar Story I

You enter a streetcar with six other passengers on the first stop of its route. On the second stop, four people come in and two get off. On the third stop, seven people come in and five get off. On the fourth stop, eight people come in and three get off. On the fifth stop, thirteen people come in and eight get off.

How old is the driver?

Did you start counting passengers in the streetcar? If you did, here is your first lesson: *Do not start solving a problem before you read it!*

Sounds obvious? Perhaps you are right. But you should not underestimate its importance. I for one underestimated some obvious things in life, and had to learn the hard way lessons like, "Do not read while you drive!"

The story above does not give us any information relevant to the age of the driver. However, relevance of information is not always obvious.

Streetcar Story II

The reunion of two friends in a streetcar sounded like this:

A. Soifer, *Mathematics as Problem Solving*, DOI: 10.1007/978-0-387-74647-0_1,
© Alexander Soifer 2009

— How are you? Thank you, I am fine.

— You just got married when we met last. Any children?

— I have three kids!

— Wow! How old are they?

— Well, if you multiply their ages, you would get 36; but if you add them up, you'd get the number of passengers in this streetcar.

— Gotcha, but you did not tell me enough to figure out their ages.

— My oldest kid is a great sportsman.

— Aha! Now I know their ages!

Find the number of passengers in the streetcar and the ages of the children.

Can the statement "my oldest kid is a great sportsman" have any relevance? It can. In fact, it does! Moreover, the fact that without this statement the second friend cannot figure out the ages of the children carries valuable information, too!

Let us take a look at the following table:

Decompositions of 36 into 3 factors x, y, z	The sum $x + y + z$
$1 \cdot 1 \cdot 36$	38
$1 \cdot 2 \cdot 18$	21
$1 \cdot 3 \cdot 12$	16
$1 \cdot 4 \cdot 9$	14
$1 \cdot 6 \cdot 6$	13
$2 \cdot 2 \cdot 9$	13
$2 \cdot 3 \cdot 6$	11
$3 \cdot 3 \cdot 4$	10

The fact that the second friend was unable to figure out the ages x, y, z of the children when he knew their sum $x + y + z$ implies that there must be at least two solutions for the given sum $x+y+z$ of ages! The table shows that only 13 appears twice in the column $x + y + z$; therefore, $x+y+z = 13$, and we know the number of passengers! We can also see the relevance of the oldest kid being a great sportsman: it rules out 1, 6, 6 and leaves 2, 2, 9!

1.2 Language

As with any other science, mathematics is formulated in an ordinary language — English in the United States. It is essential to use language correctly as well as to correctly interpret sentences. I have no intention to discuss formal mathematical language. I would like, however, to briefly talk about constructing complex sentences, and to define the meaning of "not", "and", "or", "imply," "if and only if", etc.

We will deal only with statements that are clearly true or false in a given context.

Here are a few examples of such statements:

(1) Chicago is the capital of the United States.
(2) One yard is equal to three feet.
(3) Any sports car is red.
(4) Any Ferrari is red.

As you can see, the first and third statements are false and the second statement is true. It took me a visit to my friend Bob Penkhus, a car dealer, to find out that the fourth statement is false.

The truth or falsity of a composite statement is completely determined by the truth or falsity of its components.

Negation

Given a statement A. The negation of A, denoted by $\neg A$ and read "not A," is a new statement, which is understood to assert that "A is false."

Let 1 stand for true and 0 stand for false. Then the following table defines the values of $\neg A$:

A	$\neg A$
1	0
0	1

i.e., $\neg A$ is false when A is true, and $\neg A$ is true when A is false.

Conjunction

Given statements A and B. The conjunction of A and B, denoted $A \wedge B$ and read "A and B," is a new statement which is understood to assert that "A is true and B is true." The following truth table defines $A \wedge B$:

A	B	$A \wedge B$
1	1	1
1	0	0
0	1	0
0	0	0

Disjunction

Given statements A and B. The disjunction of A and B, denoted $A \vee B$, and read "A or B," is a new statement that is understood to assert "at least one of the statements A, B is true." The following truth table defines $A \vee B$:

A	B	$A \vee B$
1	1	1
1	0	1
0	1	1
0	0	0

Implication

Given statements A and B. The implication $A \Rightarrow B$, to be read "A implies B," is a statement that is understood to assert that "if A is true, then B is true." It is defined by the following truth table:

A	B	$A \Rightarrow B$
1	1	1
1	0	0
0	1	1
0	0	1

Please note that the meaning of "implication" in mathematics is quite different from the common language use of this word: $A \Rightarrow B$ is false only if A is true and B is false.

Equivalence

Given statements A and B. The equivalence $A \Leftrightarrow B$, to be read "A equivalent B," is an abbreviation for the following statement:

$$(A \Rightarrow B) \wedge (B \Rightarrow A).$$

In order to uniquely interpret a composite statement, we some-times need to use lots of parentheses. This can make a statement quite difficult to read or evaluate. We can resolve this problem exactly the same way we do in arithmetic: by establishing the order of operations in a parenthesis-free composite statement:

$$\begin{array}{rl} \text{We apply} & \neg \quad \text{first} \\ & \wedge \quad \text{second} \\ & \vee \quad \text{third} \\ & \Rightarrow \quad \text{fourth} \\ & \Leftrightarrow \quad \text{fifth.} \end{array}$$

Finally, a composite statement that is true regardless of the truth or falsity of its components is called a *tautology*.

Problems

Prove the following tautologies:

1.1. $A \Rightarrow A$

1.2. $A \Rightarrow A \vee B$

1.3. $A \wedge B \Rightarrow A$

1.4. $\neg\neg A \Leftrightarrow A$

1.5. $\neg(A \vee B) \Leftrightarrow \neg A \wedge \neg B$ (De Morgan's Law)

1.6. $\neg(A \wedge B) \Leftrightarrow \neg A \vee \neg B$ (De Morgan's Law)

1.7. $(A \Rightarrow B) \wedge (B \Rightarrow C) \Rightarrow (A \Rightarrow C)$

1.8. $(\neg B \Rightarrow \neg A) \Leftrightarrow (A \Rightarrow B)$

1.9. $(A \wedge \neg B \Rightarrow \neg A) \Rightarrow (A \Rightarrow B)$

1.10. $(A \wedge \neg B \Rightarrow \mathcal{F}) \Rightarrow (A \Rightarrow B)$, ($\mathcal{F}$ denotes a false statement)

From now on we will use symbols:

$\land$ for "and"

$\lor$ for "or"

$\Rightarrow$ for "implies"

$\Leftrightarrow$ for "if and only if"

$\neg$ for "not"

$\exists$ for "there exists"

$\forall$ for "for every".

If $A \Rightarrow B$ is true, we say that A is a sufficient condition for B; at the same time, we say that B is a necessary condition for A. If $A \Leftrightarrow B$ is true, then B is said to be a necessary and sufficient condition for A. Please remember that a statement *converse* to $A \Rightarrow B$ is $B \Rightarrow A$. A statement *opposite* to $A \Rightarrow B$ is $\neg(A \Rightarrow B)$.

1.3 Arguing by Contradiction

Problems 1.9 and 1.10 presented the following tautologies:

$$(A \land \neg B \Rightarrow \neg A) \Rightarrow (A \Rightarrow B)$$
$$(A \land \neg B \Rightarrow \mathcal{F}) \Rightarrow (A \Rightarrow B).$$

These two tautologies justify a celebrated method of mathematical proof: arguing by contradiction.

Let us say we are given that A is true and we are asked to prove that B is true. We assume that B is not true — i.e., $\neg B$ is true — and then start with A and $\neg B$ and continue deducing until we arrive at a contradiction to what is given — i.e., at $\neg A$ — what is known to be true.

1.11. Prove the sum of a rational number and an irrational number is an irrational number.

Proof. Let r be a rational number (i.e., $r = m/n$ for some integers m, n with $n \neq 0$), and let i be an irrational number (i.e., i cannot be presented in the form s/t, where s, t are integers and $t \neq 0$).

Assume that the sum $r + i$ is a rational number, say r_1. Then if $r_1 = p/q$ for integers p, q, with $q \neq 0$, we get

$$r + i = r_1$$
$$i = r_1 - r$$
$$i = \frac{p}{q} - \frac{m}{n}$$
$$i = \frac{np - mq}{nq}.$$

That is, i is a rational number, which contradicts the given fact that i is an irrational number. Therefore, $r + i$ is irrational.   □

1.12. (*Pigeonhole Principle*) If $kn + 1$ pigeons (k, n are positive integers) are placed in n pigeonholes, then at least one of the holes contains at least $k + 1$ pigeons.

Proof. Assume that there are no holes that contain at least $k + 1$ pigeons. Then:

the 1st hole contains $\leq k$ pigeons

the 2nd hole contains $\leq k$ pigeons

$\vdots$

+ the nth hole contains $\leq k$ pigeons

total number of pigeons $\leq k \times n$ pigeons

This contradicts the given fact that there are $kn+1$ pigeons. Therefore, there is a hole that contains at least $k + 1$ pigeons.   □

Problems

1.13. Prove that the product of a nonzero rational number and an irrational number is again an irrational number.

As you probably know, a positive integer greater than 1 is called *prime* if it has exactly two divisors, 1 and itself. The Fundamental Theorem of Arithmetic states that any positive integer greater than 1 can be decomposed into the product of prime numbers and that this decomposition is unique up to the order of factors.

1.14. Given a prime p and a positive integer n. Prove that if n^2 is divisible by p, then n is divisible by p.

1.15. Prove that $\sqrt{6}$ is an irrational number.

1.16. A known theorem states that any point C of the perpendicular bisector of a segment $\overline{AB}$ is equidistant from A and B. Prove the converse.

1.17. A known theorem states that if a convex quadrilateral is inscribed in a circle, then the sums of its opposite angles are equal. Prove the converse.

1.18. A known theorem states that $m = \frac{a+b}{2}$, where m is the length of the median and a, b are the lengths of the parallel bases of a trapezoid. Prove the converse.

1.4 Pigeonhole Principle

In Section 1.3 we proved the Pigeonhole Principle, also known as the Dirichlet Principle (after its inventor, the famous mathematician Peter Gustav Dirichlet, 1805–1859). This simple principle does wonders. It is amazing how easy it is to understand this idea, and how difficult it sometimes is to discover that this idea can be applied! When you look at the problems that are solved by the Pigeonhole Principle, many of them appear to have nothing in common with each other. That is why, in April 1986, I called my lecture for the participants of the Colorado Springs Mathematical Olympiad "Invisible Pigeonhole Principle".

1.19. New York City has 7,500,000 residents. The maximal number of hairs that can grow on a human head is 500,000. Prove that there are at least 15 residents in New York City with the same number of hairs.

Solution. Let us set up 500,001 pigeonholes labeled by integers the 0 to 500,000, and put residents of New York into the holes labeled by the number of hairs on their heads. Since $7,500,000 > 14 \times 500,001 + 1$ we conclude by the Pigeonhole Principle that there is a pigeonhole with at least $14 + 1$ pigeons, i.e., there are at least 15 residents of New York with the same number of hairs.   □

1.20. (*Third Annual Colorado Springs Mathematical Olympiad, 1986*)
Santa Claus and his elves paint the plane in two colors, red and green.
Prove that there exist two points of the same color exactly one mile
apart.

Solution. Consider an equilateral triangle with each side equal to one
mile on the given plane. Since its three vertices (pigeons) are painted
in two colors (pigeonholes), we can choose two vertices painted in the
same color (at least two pigeons in a single hole).   □

1.21. (*Third Annual Colorado Springs Mathematical Olympiad, 1986*)
Given n integers, prove that either one of them is a multiple of n,
or some of them add up to a multiple of n.

Solution. Denote the n given integers by $a_1, a_2, \ldots, a_n$. Define:

$$S_1 = a_1$$
$$S_2 = a_1 + a_2$$
$$\vdots$$
$$S_n = a_1 + a_2 + \cdots + a_n.$$

If one of the numbers $S_1, S_2, \ldots, S_n$ is a multiple of n, we are done.
Assume now that none of the numbers $S_1, S_2, \ldots, S_n$ is a multiple of
n. Then all possible remainders upon the division of these numbers
by n are $1, 2, \ldots, n - 1$, i.e., we get more numbers (namely n, our
pigeons) than possible remainders ($n - 1$ possible remainders, our pigeonholes). Therefore, among the numbers $S_1, S_2, \ldots, S_n$ there exist
two numbers, say S_k and S_{k+t}, that give the same remainders upon
the division by n.

We are done, because

(1) $S_{k+t} - S_k$ is a multiple of n
(2) $S_{k+t} - S_k = a_{k+1} + a_{k+2} + \cdots + a_{k+t}$.

In other words, we found some of the given numbers, namely a_{k+1},
$a_{k+2}, \ldots, a_{k+t}$, whose sum is a multiple of n.   □

1.22. Given a real number r. Prove that among its first 99 multiples r,
$2r, \ldots, 99r$ there is at least one multiple that differs from an integer
by not more than $1/100$.

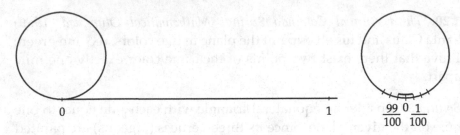

Fig. 1.1.

Solution. Let us roll the number line into a circle with circumference equal to 1 (Figure 1.1, *left*)

All integers will coincide with zero on the circle. Now we divide the circumference into 100 arcs of equal length (Figure 1.1, *right*).

If at least one multiple kr, $1 \le k \le 99$, lies on one of the arcs $[99/100, 0]$ or $[0, 1/100]$, then we are done; kr differs from an integer by no more than $1/100$.

Assume now that none of the multiples kr, $r = 1, 2, \ldots, 99$, lies on the two arcs above. We have 99 pigeons (numbers $r, 2r, \ldots, 99r$) in $100 - 2 = 98$ pigeonholes (the remaining 98 arcs). Therefore, by the Pigeonhole Principle at least two of the multiples, say kr and tr, $k > t$, lie on the same arc of length $1/100$. All that is left to notice is:

(a) $kr - tr = (k - t)r$ is one of the given 99 multiples
(b) $kr - tr$ lies on one of the arcs $[99/100, 0]$ or $[0, 1/100]$, which contradicts our assumption. □

1.23. (*A. Soifer and S.G. Slobodnik, 1973*) Forty-one rooks are placed on a 10×10 chessboard. Prove that you can choose five that do not attack each other. (We say that one rook "attacks" another if they are in the same row or column of the chessboard.)

Solution. (a) Since 41 rooks (pigeons) are placed on ten rows of the board (pigeonholes) and $41 = 4 \times 10 + 1$, there exists a row A with at least 5 rooks on it.

Let us remove row A; we now have nine rows (pigeonholes) left with at least $41 - 10 = 31$ rooks (pigeons) on them. Since $31 > 3 \times 9 + 1$, there is a row B among the nine rows with at least 4 rooks on it.

Now we remove row B. We are left with eight rows (pigeonholes) and at least $41 - 2 \times 10 = 21$ rooks (pigeons) on them. Since $21 >$

$2 \times 8 + 1$, there is a row C among the eight rows with at least 3 rooks on it.

Continuing this reasoning, we get row D with at least 2 rooks on it and row E with at least 1 rook on it.

(b) Now we are ready to select the required five rooks. First we pick any rook R_1 from row E. At least one exists there, remember!

Next, we pick a rook R_2 from row D, which is not in the same column as R_1. This can be done, too, because at least two rooks exist in row D.

Next, of course, we pick a rook R_3 from row C, which is not in the same column as R_1 and R_2. Conveniently, this can be done, since row C contains at least three rooks. Continuing this construction, we end up with five rooks $R_1, R_2, \ldots, R_5$ that are from different rows (one rook per row out of the selected rows A, B, C, D, E) and different columns; therefore, they do not attack each other. □

Problems

1.24. A three-dimensional space is painted in three colors. Prove that there are two points one mile apart painted in the same color.

1.25. Given a square of the size 1×1 and five points inside it. Prove that among the given points there are two with distance not exceeding $\frac{1}{2}\sqrt{2}$ between them.

1.26. A number of people (more than one) came to a party. Prove that at least two of them shook an equal number of hands during the party.

1.27. (*I.M. Gelfand*) Little grooves of the same width are dug across a long (very long!) straight road. The distance between the centers of any two consecutive grooves is. Prove that no matter how narrow the grooves are, a man walking along the road with a step equal to 1 sooner or later will step into a groove.

1.28. (*First Annual Southampton Mathematical Olympiad, 1986*) Grandmaster Lev Alburt plays at least one game of chess a day to keep in shape and not more than 10 games a week in order not to get too tired. Prove that if he plays long enough there will be a series of consecutive days during which he will play exactly 21 games.

1.29. (*First Annual Southampton Mathematical Olympiad, 1986*) An organization consisting of n members ($n > 5$) has $n + 1$ three-member committees, no two of which have identical membership. Prove that there are two committees that have exactly one member in common.

1.30. (*A. Soifer and S. Slobodnik, 1973*) Given 51 distinct two-digit numbers. Prove that you can choose six numbers such that any two of the six numbers have distinct digits of units and of tens.

1.31. (*A. Soifer and S. Slobodnik, 1973*) Given $r \cdot 10^{k-1} + 1$ distinct k-digit numbers, $0 < r < 9$. Prove that you can choose $r + 1$ numbers such that any two of the $r + 1$ numbers in any decimal location have distinct digits.

1.5 Mathematical Induction

Principle of Mathematical Induction I

If the first person in line is a mathematician, and every mathematician in line is followed by a mathematician, then everyone in line is a mathematician.

More seriously: *given a sequence of statements $P_1, P_2, \ldots, P_n, \ldots$.*
If

(1) P_s is true (where s is a positive integer) and
(2) for any positive integer $k \geq s$, $P_k \Rightarrow P_{k+1}$ is true,

then all of the statements in the given sequence beginning with P_s are true.

1.32. Prove the following formula for any positive integer n:

$$1 + 3 + 5 + \cdots + (2n - 1) = n^2.$$

Solution. This formula consists of a sequence of statements:

$$P_1 : 1 = 1^2$$
$$P_2 : 1 + 3 = 2^2$$
$$P_3 : 1 + 3 + 5 = 3^2$$
$$\vdots$$
$$P_n : 1 + 3 + 5 + \cdots + (2n - 1) = n^2.$$
$$\vdots$$

(1) P_1 is certainly true.

(2) Assume P_k is true, i.e.,

$$1 + 3 + \cdots + (2k - 1) = k^2.$$

Adding $2k + 1$ to both sides of this equality, we get

$$1 + 3 + 5 + \cdots + (2k - 1) + (2k + 1) = k^2 + 2k + 1,$$

but

$$k^2 + 2k + 1 = (k + 1)^2.$$

Therefore,

$$1 + 3 + 5 + \cdots + (2k + 1) = (k + 1)^2,$$

i.e., P_{k+1} is true.

We proved that the implication $P_k \Rightarrow P_{k+1}$ is true for any positive integer k.

By the Principle of Mathematical Induction, P_n is true for every positive integer n. The formula is proven. □

I cannot resist showing here one more solution to this problem, a geometric one; see Figure 1.2.

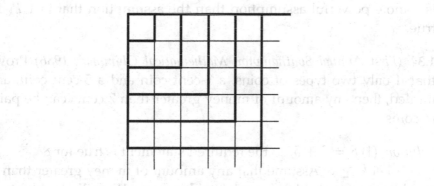

Fig. 1.2.

Solve the next problem on your own.

1.33. Prove that for any positive integer n,

$$1 + 2 + \cdots + n = \frac{n(n+1)}{2}.$$

Please note that condition (2) in the Principle of Mathematical Induction only requires that the implication $P_k \Rightarrow P_{k+1}$ be true. It does not require that P_{k+1} be true. In fact, a sequence $P_1, P_2, \ldots, P_n, \ldots$ of all false statements satisfies condition (2)!

Thus, condition (1), which requires that P_s be true, is an essential part of the Principle.

Another version of the Principle of Mathematical Induction, equivalent to the original one, but giving us a somewhat more powerful tool is the following:

Principle of Mathematical Induction II

Given a sequence of statements $P_1, P_2, \ldots, P_n, \ldots$. If

(1) P_s is true (where s is a positive integer) and
(2) $P_s \wedge P_{s+1} \wedge \cdots \wedge P_k \Rightarrow P_{k+1}$ is true for any positive integer $k \geq s$,

then all of the statements in the given sequence beginning with P_s are true.

The difference between this version and the original one is that we can assume that the statements $P_s, P_{s+1}, \ldots, P_k$ are all true and prove then that P_{k+1} is true as well. And, of course, in some cases this is a more powerful assumption than the assumption that just P_k is true.

1.34. (*First Annual Southampton Mathematical Olympiad, 1986*) Prove that if only two types of coins, a 3-cent coin and a 5-cent coin, are minted, then any amount of money greater than 7 cents can be paid in coins.

Solution. (1) $8 = 3 + 5$, so the required statement is true for 8.

(2) Let $k \geq 8$. Assume that any amount of money greater than 7 and less than or equal to k can be paid in coins. We will consider two cases.

1: If $k - 5 \geq 8$, then by assumption $k - 5$ can be paid in coins, and so can $k + 1$:

$$k + 1 = (k - 5) + 3 + 3.$$

2: If $k - 5 < 8$, then $k + 1 < 14$, and we have to check just a few values, namely $k + 1 = 9, 10, 11, 12, 13$. But

$$9 = 3 + 3 + 3$$
$$10 = 5 + 5$$
$$11 = 3 + 3 + 5$$
$$12 = 3 + 3 + 3 + 3$$
$$13 = 3 + 5 + 5.$$

We proved that in either case $k + 1$ can be paid in coins. Thus, any amount of money greater than 7 cents can be paid in coins. □

The limitation of the method of mathematical induction is that it can help us verify a hypothesis we have but offers no help in coming up with a hypothesis. For the latter we have to use our intuition often coupled with experimentation.

1.35. We say that several straight lines on the plane are in general position if no two lines are parallel and no three lines have a point in common. In how many regions do n straight lines in general position partition the plane?

Solution. Let us denote the number of regions n lines in general position partition the plane into by $S(n)$. Now let us experiment; we draw one line on the plane and get $S(1) = 2$; we add one more line to see that $S(2) = 4$; one more line will show that $S(3) = 7$; one more line will show that $S(4) = 11$ (see Figure 1.3).

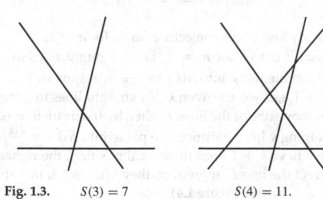

Fig. 1.3. $S(3) = 7$ $S(4) = 11$.

Let us put the data in a table:

Number of Lines n	$S(n)$	Difference $S(n) - S(n-1)$
1	2	
2	4	2
3	7	3
4	11	4

We notice that

$$S(n) = S(n-1) + n.$$

This strikingly resembles the relation of Problem 1.33. Remember, $1 + 2 + \cdots + n = \frac{n(n+1)}{2}$, i.e., if $S_1(n)$ denotes $1 + 2 + \cdots + n$, we get the same relation $S_1(n) = S_1(n-1) + n$.

So let us check the hypothesis $S(n) = \frac{n(n+1)}{2}$:

n	$S(n)$	$\dfrac{n(n+1)}{2}$
1	2	1
2	4	3
3	7	6
4	11	10

Our hypothesis does not work, but we can now see from the table above that $S(n)$ and $\frac{n(n+1)}{2}$ differ only by 1, always by 1! Thus, we can conjecture:

$$S(n) = \frac{n(n+1)}{2} + 1.$$

As we already know, our conjecture holds for $n = 1$.

Assume that it is true for $n = k$, i.e., k straight lines in general position partition the plane into $S(k) = \frac{k(k+1)}{2} + 1$ areas.

Let $n = k + 1$, i.e., we are given $k + 1$ straight lines in general position. If we remove one of the lines L, then by the inductive assumption, the remaining k lines partition the plane into $S(k) = \frac{k(k+1)}{2} + 1$ areas. Since we have $k + 1$ lines in general position, the remaining k lines all intersect the line L; moreover, they intersect L in k different points $a_1, a_2, \ldots, a_k$ (see Figure 1.4).

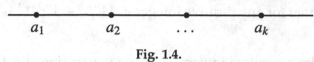

Fig. 1.4.

These k points split the line L into $k + 1$ intervals. Each of these intervals splits one region of the partition of the plane by k lines into two new areas. That is, instead of $k + 1$ old areas we get $2(k+1)$ new areas:

$$S(k + 1) = S(k) + (k + 1).$$

Therefore,

$$S(k + 1) = \frac{k(k + 1)}{2} + 1 + (k + 1) = \frac{(k + 1)(k + 2)}{2} + 1.$$

In other words, our conjecture holds for $n = k+1$. Thus, n straight lines in general position partition the plane into $\frac{n(n+1)}{2} + 1$ regions. $\square$

Problems

1.36. Prove that for any positive integer n:

$$1^2 + 2^2 + \cdots + n^2 = \frac{n(n + 1)(2n + 1)}{6}.$$

1.37. Prove that for any positive integer n:

$$1^3 + 2^3 + \cdots + n^3 = \frac{n^2(n + 1)^2}{4}.$$

Note that coupled with Problem 1.33, the equality 1.37 gives us a beautiful corollary:

$$1^3 + 2^3 + \cdots + n^3 = (1 + 2 + \cdots + n)^2.$$

1.38. Prove that for any positive integer n,

$$\frac{1}{1 \times 2} + \frac{1}{2 \times 3} + \cdots + \frac{1}{n(n + 1)} = \frac{n}{n + 1}.$$

1.39. Prove that for any positive integer n, $n^3 - n$ is divisible by 6.

1.40. Prove that for any positive integer n, $11^{n+2} + 12^{2n+1}$ is divisible by 133.

1.41. Let n be a positive integer. A plane is partitioned into several regions by n lines. Prove that the plane can be colored in two colors in such a way that any two regions with a common border (more than one point) are colored in different colors.

2

Numbers

2.1 Integers

You remember the definition of a prime number. On p. 7, we defined a prime number and formulated the Fundamental Theorem of Arithmetic. Numerous beautiful results can be presented here, but I will limit myself to problems illustrating some ideas that require practically no knowledge of number theory.

2.1. Prove that for any integer n, $n^5 - 5n^3 + 4n$ is divisible by 120.

Solution. First of all, let us decompose $P(n) = n^5 - 5n^3 + 4n$ and 120 into factors:

$$P(n) = n(n^4 - 5n^2 + 4)$$
$$= n(n^2 - 1)(n^2 - 4)$$
$$= (n - 2)(n - 1)n(n + 1)(n + 2);$$

$$120 = 2^3 \cdot 3 \cdot 5.$$

Since for any integer n, $P(n)$ is a product of five consecutive integers, and one of any five consecutive integers is a multiple of 5, $P(n)$ is divisible by 5.

Similarly, out of any three consecutive integers, one is a multiple of 3; therefore, $P(n)$ is divisible by 3 for any integer n.

Out of any four consecutive integers, one is a multiple of 4, plus one more is even. Therefore, $P(n)$ is divisible by $4 \times 2 = 8$ for any integer n.

A. Soifer, *Mathematics as Problem Solving*, DOI: 10.1007/978-0-387-74647-0_2,
© Alexander Soifer 2009

Due to the Fundamental Theorem of Arithmetic, $P(n)$ is divisible by $23 \times 3 \times 5 = 120$. $\square$

As a 7th-grader, I faced this problem at the Moscow Mathematical Olympiad in the spring of 1962, which offered us four problems and four hours to solve them. It took me no time to solve the other three problems. After much effort, I finally conquered this problem. The divisibility of the expression by 8 was, of course, the difficulty to overcome.

Given a quadratic equation $ax^2 + bx + c = 0$; the number $D = b^2 - 4ac$ is called the *discriminant* of the equation.

2.2. Is there an integer x such that $x^2 + x + 3$ is a multiple of 121?

Solution I Assume that

$$x^2 + x + 3 = 121k,$$

where x and k are integers. We then have a quadratic equation in x:

$$x^2 + x + (3 - 121k) = 0.$$

In order for a solution to be an integer (remember, the problem asks whether an integer x exists!), the discriminant of the equation has to be a perfect square:

$$4 \times 121k - 11 = n^2$$

where n is an integer; that is,

$$n^2 = 11(4k \times 11 - 1).$$

This means that n^2 is divisible by 11 but not by 11^2. On the other hand, due to Problem 1.14, since n^2 is divisible by 11, n is divisible by 11 as well, which in turn implies that n^2 is divisible by 11^2. This is a contradiction.

Therefore, there is no integer x such that $x^2 + x + 3$ is a multiple of 121. $\square$

Solution II $(x^2 + x + 3)$ is divisible by 121 if and only if $4(x^2 + x + 3)$ is divisible by 121, but

$$4(x^2 + x + 3) = (2x + 1)^2 + 11.$$

If for some integers x and k,

$$(2x + 1)^2 + 11 = 11^2 \times k,$$

then

$$(2x + 1)^2 = 11(11k - 1).$$

Just as in the first solution, the contradiction is derived from the fact that a square, namely $(2x + 1)^2$, is divisible by 11 but is not divisible by 11^2.   □

2.3. Find all integral solutions of the equation

$$x^2 + y^2 + x + y = 3.$$

Solution I Let $L(x, y) = x^2 + y^2 + x + y = x(x+1) + y(y+1)$. For any integer x, $x(x+1)$ is even as the product of two consecutive integers. Similarly, $y(y+1)$ is even. Thus for any integers x, y, $L(x, y)$ is even and therefore not equal to 3.

The solution set is empty.   □

Solution II By multiplying both sides of the given equation by 4, we get:

$$(4x^2 + 4x) + (4y^2 + 4y) = 12,$$

or

$$(2x + 1)^2 + (2y + 1)^2 = 14.$$

On the other hand, a direct check shows that 14 is not the sum of two squares of integers.   □

Problems

2.4. Prove that for any integer n, $n^5 - n$ is divisible by 30.

2.5. Prove that for a prime p greater than 3, $p^2 - 1$ is divisible by 24.

2.6. Prove that for any primes p and q, each greater than 3, $p^2 - q^2$ is divisible by 24.

2.7. Can $4p + 1$ be a prime number if both p and $2p + 1$ are primes and $p > 3$?

2.8. Prove that the remainder upon dividing any prime number by 30 is again a prime.

2.9. Find the integer solutions to

$$15x^2 - 7y^2 = 9.$$

2.10. Prove that for any positive integer n, $10^n + 18n - 1$ is divisible by 27.

2.2 Rational and Irrational Numbers

We have already met rational and irrational numbers in Chapter 1, Section 1.3 (Problems 1.11, 1.13, and 1.15). As you know, rational numbers can be presented in the form m/n, where m and n are integers and $n \neq 0$. But how do we recognize whether a number given as a decimal fraction is rational or irrational? Rational numbers are terminal or infinite repeating decimal fractions.

2.11. Prove that the number

$$A = 0.101001000\ldots,$$

where the number of zeros between units increases by one, is irrational.

Solution. Assume that A is a repeating fraction, i.e., after the first k digits, the same sequence of n digits (we'll call it *period*) repeats. Since the number of consecutive zeros in the decimal representation of A is increasing, we can find $2n + k$ consecutive zeros, but this implies that all n digits of the period are zeros. Therefore, in the decimal decomposition of A we get only zeros from some point on.

However, this contradicts the definition of A, which allows us to find a digit one further right than any given digit of the decimal representation of A.   □

2.12. The numbers a, b, and $\sqrt{a} + \sqrt{b}$ are rational. Prove that the numbers $\sqrt{a}$ and $\sqrt{b}$ are rational as well.

Solution. The numbers a and b are rational; therefore, $(a + b)$ is rational. The numbers $(a + b)$ and $(\sqrt{a} + \sqrt{b})$ are rational, thus $\sqrt{a} - \sqrt{b} = \frac{a+b}{\sqrt{a}+\sqrt{b}}$ is rational. Now we can see that

$$\sqrt{a} = \tfrac{1}{2}[(\sqrt{a} + \sqrt{b}) + (\sqrt{a} - \sqrt{b})]$$

is rational; so is $\sqrt{b} = (\sqrt{a} + \sqrt{b}) - \sqrt{a}$.   □

2.13. Prove that $1 + \sqrt{5}$ cannot be written as a sum of squares of numbers of the form $a + b\sqrt{5}$ with rational a and b.

Solution. Let us first note that if for integers x_1, x_2, y_1, y_2,

$$x_1 + y_1\sqrt{5} = x_2 + y_2\sqrt{5},$$

then $x_1 = x_2$ and $y_1 = y_2$. Indeed, otherwise we would get $x_1 \neq x_2$ and $y_1 \neq y_2$, $\sqrt{5} = \frac{x_1 - x_2}{y_2 - y_1}$, with an irrational left side and a rational right side.

Now let us assume that

$$1 + \sqrt{5} = (a_1 + b_1\sqrt{5})^2 + (a_2 + b_2\sqrt{5})^2 + \cdots + (a_n + b_n\sqrt{5})^2.$$

Due to the uniqueness proven above, we can conclude that

$$1 - \sqrt{5} = (a_1 - b_1\sqrt{5})^2 + (a_2 - b_2\sqrt{5})^2 + \cdots + (a_n - b_n\sqrt{5})^2.$$

But $1 - \sqrt{5} < 0$, while the right side is nonnegative (as a sum of squares) — a contradiction.   □

2.14. Let p/q, where p and q are integers and their greatest common divisor is 1, be a solution of the algebraic equation

$$a_0 x^n + a_1 x^{n-1} + \cdots + a_n = 0$$

with integral coefficients a_i. Prove that p is a divisor of a_n and q is a divisor of a_0.

Solution. Essentially, we are given the equality

$$a_0 \frac{p^n}{q^n} + a_1 \frac{p^{n-1}}{q^{n-1}} + \cdots + a_{n-1} \frac{p}{q} + a_n = 0.$$

Therefore,

$$a_0 p^n = q(-a_1 p^{n-1} - \cdots - a_n q^{n-1}),$$

i.e., q is a divisor of $a_0 p^n$. Since $\gcd(p, q) = 1$, this implies that q is a divisor of a_0.

Similarly,

$$a_n q^n = p(-a_{n-1} q^{n-1} - \cdots - a_0 p^{n-1}),$$

and thus p is a divisor of a_n. $\qquad\square$

The statement of Problem 2.14 has a very important consequence:

Corollary 2.1. *Any rational solution of the equation*

$$x^n + a_1 x^{n-1} + \cdots + a_n = 0$$

with integral coefficients is an integer.

Prove it!

Problems

2.15. The number A is given as a decimal fraction:

$$A = 0.10000000001 \ldots ,$$

where units occupy the first, tenth, hundredth, thousandth, etc., positions after the dot, and zeros everywhere else. Prove that

(a) A is an irrational number;
(b) A^2 is an irrational number.

2.16. Prove that for any integer n,

$$\frac{n}{3} + \frac{n^2}{2} + \frac{n^3}{6}$$

is an integer.

2.17. Prove that for any positive integer n,

$$\frac{n}{6} + \frac{n^2}{2} + \frac{n^3}{3}$$

is an integer.

2.18. Prove that $\sqrt[3]{2}$ cannot be written in the form $p + q\sqrt{r}$, where p, q, and r are rational numbers.

2.19. Solve Problem 1.38 without the use of mathematical induction, i.e., prove that for any positive integer n,

$$\frac{1}{1 \times 2} + \frac{1}{2 \times 3} + \cdots + \frac{1}{n(n+1)} = \frac{n}{n+1}.$$

3

Algebra

3.1 Proof of Equalities and Inequalities

3.1. Prove that for any real numbers a, b, c, the sum $a + b + c$ is a divisor of

$$a^3 + b^3 + c^3 - 3abc.$$

Solution. By using the equality

$$x^3 + y^3 = (x + y)^3 - 3xy(x + y)$$

twice, we get:

$$
\begin{aligned}
a^3 + b^3 + c^3 - 3abc \\
&= (a + b)^3 + c^3 - 3ab(a + b) - 3abc \\
&= (a + b + c)^3 - 3(a + b)c(a + b + c) - 3ab(a + b + c) \\
&= (a + b + c)(a^2 + b^2 + c^2 - ab - ac - bc).
\end{aligned}
$$
$\square$

3.2. What is wrong with the following "proof" of the inequality $\frac{a+b}{2} \geq \sqrt{ab}$:

$$\frac{a + b}{2} \geq \sqrt{ab} \Rightarrow \frac{(a + b)^2}{4} \geq ab$$

$$\Rightarrow a^2 + 2ab + b^2 \geq 4ab$$

$$\Rightarrow (a - b)^2 \geq 0.$$

A. Soifer, *Mathematics as Problem Solving*, DOI: 10.1007/978-0-387-74647-0_3,
© Alexander Soifer 2009

The last inequality is true, therefore

$$\frac{a+b}{2} \geq \sqrt{ab}.$$

Solution. Something must be wrong, because the inequality to be proved is false for, say, negative a and b, and it is not defined when one of the numbers a, b is positive and one negative. All of the implications in our chain are true, but the fact that we deduced a true inequality from the one to be proved has nothing to do with proving that inequality. Well, almost nothing. This chain can serve as *analysis*, which can help find a proof, but the proof must be a chain of implications, which starts with the inequality known to be true and ends with the required inequality.   □

3.3. Prove that for any nonnegative numbers a, b,

$$\frac{a+b}{2} \geq \sqrt{ab}.$$

Solution I (a) Analysis:

$$\frac{a+b}{2} \geq \sqrt{ab} \Rightarrow a - 2\sqrt{ab} + b \geq 0$$
$$\Rightarrow \left(\sqrt{a} - \sqrt{b}\right)^2 \geq 0.$$

(b) Proof:

$$\left(\sqrt{a} - \sqrt{b}\right)^2 \geq 0$$

for any nonnegative a, b, for the left side is a square of a real number,

$$\Rightarrow a - 2\sqrt{ab} + b \geq 0$$
$$\Rightarrow \frac{a+b}{2} \geq \sqrt{ab}.$$ □

Instead of tracing two chains of implications, we can make sure that every implication in the original chain (analysis) is reversible.

Solution II

$$\frac{a+b}{2} \geq \sqrt{ab} \Leftrightarrow a - 2\sqrt{ab} + b \geq 0$$

$$\Leftrightarrow \left(\sqrt{a} - \sqrt{b}\right)^2 \geq 0,$$

which is true for any nonnegative a and b.

Note that the equality is achieved if and only if $a = b$. □

3.4. Prove that for any nonnegative x, y, z,

$$x^2 + y^2 + z^2 \geq xy + yz + xz.$$

Solution. Using the inequality proved in Problem 3.3, we get:

$$\frac{x^2 + y^2}{2} \geq xy$$

$$\frac{y^2 + z^2}{2} \geq yz$$

$$\frac{x^2 + z^2}{2} \geq xz.$$

All that is left to do is to add up these three inequalities! □

3.5. Prove that for any nonnegative a, b, c,

$$\frac{a+b+c}{3} \geq \sqrt[3]{abc}.$$

Solution. From Problem 3.1, we know that

$$x^3 + y^3 + z^3 - 3xyz = (x + y + z)(x^2 + y^2 + z^2 - xy - yz - xz).$$

Due to the inequality proved in Problem 3.4,

$$x^2 + y^2 + z^2 - xy - yz - xz \geq 0.$$

Combining these two facts, we conclude that for any nonnegative x, y, z,

$$x^3 + y^3 + z^3 - 3xyz \geq 0.$$

All that is left to do is to let $x^3 = a$, $y^3 = b$, $y^3 = c$ to get the required inequality

$$\frac{a+b+c}{3} \geq \sqrt[3]{abc}.$$ □

3.6. Prove that for any positive integer n and nonnegative $a_1, a_2, \ldots,$ a_n,

$$\frac{a_1 + a_2 + \cdots + a_n}{n} \geq \sqrt[n]{a_1 a_2 \cdots a_n}.$$

Moreover, the equality is achieved if and only if $a_1 = a_2 = \cdots = a_n$.

Solution. (a) We will first prove by induction on m that this inequality is true for $n = 2^m$. It is true for $m = 1$ (we proved it in Problem 3.3).

Assume that the inequality is true for $p = 2^k$, i.e., for any nonnegative $a_1, a_2, \ldots, a_p$,

$$\frac{a_1 + a_2 + \cdots + a_p}{p} \geq \sqrt[p]{a_1 a_2 \cdots a_p}.$$

Now let $a_1, a_2, \ldots, a_{2p}$ be nonnegative numbers. By using first the inductive assumption and then the inequality of Problem 3.3, we get:

$$\frac{a_1 + a_2 + \cdots + a_{2p}}{2p} = \frac{a_1 + a_2}{2p} + \frac{a_3 + a_4}{2p} + \cdots + \frac{a_{2p-1} + a_{2p}}{2p}$$

$$\geq \sqrt[p]{\frac{a_1 + a_2}{2} \frac{a_3 + a_4}{2} \cdots \frac{a_{2p-1} + a_{2p}}{2}}$$

$$\geq \sqrt[p]{\sqrt{a_1 a_2}\sqrt{a_3 a_4} \cdots \sqrt{a_{2p-1} a_{2p}}}$$

$$= \sqrt[2p]{a_1 a_2 \cdots a_{2p}}.$$

(b) Now we can prove the required inequality for any positive integer n. Indeed, if $n = 2^m$ for some m, the inequality is proved by (a); otherwise, there exist positive integers t and m such that $n + t = 2^m$. For any nonnegative $a_1, a_2, \ldots, a_{n+t}$,

$$\frac{a_1 + a_2 + \cdots + a_n + a_{n+1} + \cdots + a_{n+t}}{n + t}$$

$$\geq \sqrt[n+t]{a_1 a_2 \cdots a_n a_{n+1} \cdots a_{n+t}}.$$

All that is left to do is to define

$$a_{n+1} = a_{n+2} = \cdots = a_{n+1} = \frac{a_1 + a_2 + \cdots + a_n}{n}$$

and plug it into the inequality above:

$$\frac{a_1 + a_2 + \cdots + a_n + t \cdot \frac{a_1 + a_2 + \cdots + a_n}{n}}{n + t}$$

$$\geq \sqrt[n+t]{a_1 a_2 \cdots a_n \left(\frac{a_1 + a_2 + \cdots + a_n}{n}\right)^t}.$$

If we denote $\frac{a_1 + a_2 + \cdots + a_n}{n} = A$ and $\sqrt[n]{a_1 a_2 \cdots a_n} = B$, the inequality above can be rewritten as

$$\frac{nA + tA}{n + t} \geq \sqrt[n+t]{B^n A^t},$$

that is,

$$A \geq \sqrt[n+t]{B^n A^t}.$$

Therefore,

$$A^{n+t} \geq B^n A^t,$$

or

$$A^n \geq B^n,$$

and finally

$$A \geq B.$$

But this is exactly the required inequality!

(c) If $a_1 = a_2 = \cdots = a_n \geq 0$, then we get the equality. Assume now that not all nonnegative numbers $a_1, a_2, \ldots, a_n$ are equal. Without loss of generality, we can assume that $a_1 \neq a_2$. Then

$$\frac{a_1 + a_2}{2} > \sqrt{a_1 a_2}, \text{ or } \frac{(a_1 + a_2)^2}{2} > a_1 a_2,$$

and we get

$$\frac{a_1 + a_2 + \cdots + a_n}{n} = \frac{\frac{a_1 + a_2}{2} + \frac{a_1 + a_2}{2} + a_3 + \cdots + a_n}{n}$$

$$\geq \sqrt[n]{\left(\frac{a_1 + a_2}{2}\right)^2 a_3 \cdots a_n} > \sqrt[n]{a_1 a_2 \cdots a_n}.$$

Thus the equality takes place if and only if $a_1 = a_2 = \cdots = a_n$. □

3.7. Prove that for any positive integer $n \geq 2$,

$$\frac{1}{2^2} + \frac{1}{3^2} + \cdots + \frac{1}{n^2} < 1.$$

Solution. Clearly,

$$\frac{1}{2^2} + \frac{1}{3^2} + \cdots + \frac{1}{n^2} < \frac{1}{1 \cdot 2} + \frac{1}{2 \cdot 3} + \cdots + \frac{1}{(n-1)n}.$$

Now, if you solved Problem 1.38 or 2.19 you know that

$$\frac{1}{1 \cdot 2} + \frac{1}{2 \cdot 3} + \cdots + \frac{1}{(n-1)n} < \frac{n-1}{n}.$$

Therefore,

$$\frac{1}{2^2} + \frac{1}{3^2} + \cdots + \frac{1}{n^2} < \frac{n-1}{n} < 1.$$

But what if you could not solve Problem 2.19? Here is a solution for you:

$$\frac{1}{1 \cdot 2} + \frac{1}{2 \cdot 3} + \cdots + \frac{1}{(n-1)n} = \left(\frac{1}{1} - \frac{1}{2}\right) + \left(\frac{1}{2} - \frac{1}{3}\right) + \cdots + \left(\frac{1}{n-1} - \frac{1}{n}\right).$$

As you can see, all the terms except for the first and last ones cancel out, and we get the required

$$1 - \frac{1}{n} = \frac{n-1}{n}. \qquad \square$$

Problems

3.8. Prove that for any positive numbers a and b,

$$\frac{2}{\dfrac{1}{a} + \dfrac{1}{b}} \leq \sqrt{ab} \leq \frac{a+b}{2} \leq \sqrt{\frac{a^2 + b^2}{2}}.$$

Moreover, the equalities are true if and only if $a = b$.

3.9. Prove that for any nonnegative numbers a and b,

$$\frac{a+b}{2} \leq \sqrt[3]{\frac{a^3 + b^3}{2}}.$$

3.10. Prove that for any nonnegative x, y, z,

$$(x + y)(y + z)(x + z) \geq 8xyz.$$

3.11. Prove that for any $a_1, a_2, \ldots, a_n$,

$$\frac{a_1 + a_2 + \cdots + a_n}{n} \leq \sqrt{\frac{a_1^2 + a_2^2 + \cdots + a_n^2}{n}}.$$

3.12. Prove that for any positive integer n,

$$\frac{1}{3^2} + \frac{1}{5^2} + \cdots + \frac{1}{(2n + 1)^2} < \frac{1}{4}.$$

3.13. Prove that there is a better estimate than the one we proved in Problem 3.7: for any positive integer $n \geq 2$,

$$\frac{1}{2^2} + \frac{1}{3^2} + \cdots + \frac{1}{n^2} < \frac{2}{3}.$$

3.2 Equations, Inequalities, Their Systems, and How to Solve Them

The wealth of material available for this section is so great that it alone could fill a book, or even books. Without any claim of completeness, I want to present here a few ideas. Unless otherwise stated, we will solve equations and inequalities in the set of real numbers.

3.14. Solve the following equation:

$$(x^2 - x + 1)(x^2 - x + 2) = 12.$$

Solution. Let $x^2 - x + 1 = y$ and express the given equation in terms of y:

$$y(y + 1) = 12,$$

that is,

$$y^2 + y - 12 = 0.$$

Now, we can solve this quadratic equation, with

$$y_1 = -4$$
$$y_2 = 3$$

Rewriting it back in terms of x, we get:

$x^2 - x + 1 = -4$	$x^2 - x + 1 = 3$
$\Rightarrow \quad x^2 - x + 5 = 0$	$x^2 - x - 2 = 0$
no solutions	$x_1 = -1$ or $x_2 = 2$
in real numbers	

Thus the solution set is $\{-1, 2\}$. □

The reader is probably familiar with the Viète theorem:

Theorem 3.1 (Viète). *The solution set $\{x_1, y_1\}$ of the quadratic equation*

$$z^2 + pz + q = 0$$

satisfies the following system of equalities:

$$x_1 + y_1 = -p$$
$$x_1 y_1 = q.$$

It is sometimes very helpful to know that the converse of the Viète theorem is true, too:

Theorem 3.2. *If the ordered pair (x_1, y_1) is a solution of the system*

$$x + y = -p$$
$$xy = q,$$

then the set $\{x_1, y_1\}$ is the solution set of the equation

$$z^2 + pz + q = 0.$$

Prove it!

3.15. Solve the following system of equations:

$$x^3 + y^3 = -7$$
$$xy = -2.$$

Solution. Denote

$$x + y = -p$$
$$xy = q.$$

If we can find p and q, then by the converse of the Viète theorem, $\{x, y\}$ will be the solution set of the equation

$$z^2 + pz + q = 0.$$

Let us therefore rewrite the given system in terms of p and q. Because $x^3 + y^3 = (x + y)^3 - 3xy(x + y)$, it is easy to do:

$$(-p)^3 - 3q(-p) = -7$$
$$q = -2,$$

that is,

$$p^3 - 3pq = 7$$
$$q = -2.$$

By substituting -2 for q in the first equation, we get

$$p^3 + 6p - 7 = 0$$

that is,

$$(p - 1)(p^2 + p + 7) = 0$$

$p - 1 = 0$	$p^2 + p + 7 = 0$
$p = 1$	no real solutions

We have $\left\{ \begin{smallmatrix} p=1 \\ q=-2 \end{smallmatrix} \right.$, therefore $\{x, y\}$ is the solution set of the equation $z^2 + z - 2 = 0$. Solving this equation, we get $z_1 = 1$, $z_2 = -2$; therefore, the original system has two solutions:

$$\begin{cases} x = 1 \\ y = -2 \end{cases}, \qquad \begin{cases} x = -2 \\ y = 1 \end{cases} \qquad \square$$

Quite often, in order to solve a given system of equations, we replace it with one equation. It is much more surprising that *sometimes replacing the given equation by a system of equations proves to be very productive*, too.

3.16. Solve the following equation:

$$x^2 + \left(\frac{x}{x+1}\right)^2 = 1.^1$$

Solution. Straightforward simplification will bring you (try it!) to a pretty hopeless equation:

$$x^4 + 2x^3 + x^2 - 2x - 1 = 0.$$

(Of course, there is a formula for solving fourth-degree equations, but I have yet to meet a person who remembers it by heart!)

Let $\frac{x}{x+1} = y$. Our equation is equivalent to the following system of equations:

$$\begin{cases} x^2 + y^2 = 1 \\ \dfrac{x}{x+1} = y \end{cases}$$

or

$$x^2 + y^2 = 1$$
$$x - y = xy.$$

Denote

$$x + (-y) = -p$$
$$x(-y) = q.$$

Since $x^2 + y^2 = [x + (-y)]^2 - 2x(-y)$, this system can be rewritten in terms of p and q as follows:

$$(-p)^2 - 2q = 1$$
$$-p = -q.$$

Our goal now is to find p and q, and then to use the converse of the Viète theorem to find x and $-y$.

We have

$$\begin{cases} p^2 - 2q = 1 \\ p = q, \end{cases}$$

---

[1] Remark for the second edition: Cecil Rousseau found a way to solve this equation by combining terms and rewriting it as $(x^2 + x + 1)^2 - 2(x + 1)^2 = 0$.

therefore,

$$p^2 - 2p - 1 = 0.$$

Thus,

$$\begin{cases} p_1 = 1 + \sqrt{2} \\ q_1 = 1 + \sqrt{2} \end{cases}, \qquad \begin{cases} p_2 = 1 - \sqrt{2} \\ q_2 = 1 - \sqrt{2} \end{cases}$$

and $\{x, -y\}$ is the solution set of the equation

$$z^2 + pz + q = 0.$$

Accordingly, we have two cases:

1: $p = q = 1 + \sqrt{2}.$

$$z^2 + (1 + \sqrt{2})z + (1 + \sqrt{2}) = 0.$$

Since the discriminant

$$D = 1 + \sqrt{2}^2 - 4(1 + \sqrt{2}) < 0,$$

there are no real solutions.

2: $p = q = 1 - \sqrt{2}.$

$$z^2 + (1 + \sqrt{2})z + (1 - \sqrt{2}) = 0$$

$$z_{1,2} = \frac{\sqrt{2} - 1 \pm \sqrt{2\sqrt{2} - 1}}{2}.$$

Therefore, we get two solutions:

$$\begin{cases} x = z_1 \\ -y = z_2, \end{cases} \qquad \begin{cases} x = z_2 \\ -y = z_1. \end{cases}$$

Since we are not really interested in solutions for y (after all, y was not a part of the original problem, it was our creation!), we finally get:

$$x_1 = \tfrac{1}{2}\left(\sqrt{2} - 1 + \sqrt{2\sqrt{2} - 1}\right),$$

$$x_2 = \tfrac{1}{2}\left(\sqrt{2} - 1 - \sqrt{2\sqrt{2} - 1}\right) \qquad \square$$

3.17. Solve the following equation:

$$x^{x^2-5x+6} = 1.$$

Solution. By taking the natural logarithm of both sides of the equation, we get

$$(x^2 - 5x + 6) \ln |x| = 0.$$

(*Note that in order not to lose any solutions, we must use the absolute value sign. As a result, we might gain some extraneous solutions. Therefore all solutions x such that x < 0, will have to be checked.*)

$$
\begin{array}{c|c}
x^2 - 5x + 6 = 0 & \ln |x| = 0 \\
x_1 = 2;\ x_2 = 3 & x_3 = 1;\ x_4 = -1.
\end{array}
$$

The check shows that $x_4 = -1$ is a solution of this equation, as well, of course, as $x = 1, 2, 3$. ☐

3.18. Solve the following equation:

$$\sqrt{x+1} - \sqrt{x-1} = 1.$$

Solution. By multiplying both sides of the given equation by $\sqrt{x+1} + \sqrt{x-1}$, we get

$$\sqrt{x+1} + \sqrt{x-1} = 2.$$

Now we add this equation to the original one to get

$$2\sqrt{x+1} = 3.$$

Therefore,

$$x + 1 = \tfrac{9}{4}$$

or

$$x = \tfrac{5}{4}.$$ ☐

3.19. Solve the following equation:

$$\sqrt[3]{1 + \sqrt{x}} + \sqrt[3]{1 - \sqrt{x}} = \sqrt[3]{5}.$$

Solution. Keeping in mind that

$$(a+b)^3 = a^3 + b^3 + 3ab(a+b),$$

we compute the cubes of both sides of the given equation:

$$2 + 3\sqrt[3]{1-x}\left(\sqrt[3]{1+\sqrt{x}} + \sqrt[3]{1-\sqrt{x}}\right) = 5.$$

But $\sqrt[3]{1+\sqrt{x}} + \sqrt[3]{1-\sqrt{x}} = \sqrt[3]{5}$; therefore, we get

$$2 + 3\sqrt[3]{1-x} \cdot \sqrt[3]{5} = 5,$$

that is,

$$1 - x = \tfrac{1}{5},$$

$$x = \tfrac{4}{5}.$$

Note: When we substitute the right side of the original equation into the left side of any of the subsequent equations, we will not lose a solution, but we could gain one! Therefore, we must check our solutions. The solution we have obtained checks out to be a true solution. To see the importance of this remark, try to solve the same problem with -1 on the right side instead of $\sqrt[3]{5}$. You will get $x = 0$, which does not check out!   □

3.20. Solve the following inequality:

$$\sqrt{\frac{x^3 + 8}{x}} > x - 2.$$

Solution. First we determine the domain of the given inequality:

$$\begin{cases} x^3 + 8 \geq 0 \\ x > 0 \end{cases} \Rightarrow \begin{cases} x \geq -2 \\ x > 0 \end{cases} \Rightarrow x > 0$$

or

$$\begin{cases} x^3 + 8 \leq 0 \\ x < 0 \end{cases} \Rightarrow \begin{cases} x \leq -2 \\ x < 0 \end{cases} \Rightarrow x \leq -2.$$

Thus the domain is $(-\infty, -2) \cup (0, +\infty)$. We will consider two cases:

1: $x - 2 < 0$, or $x < 2$. In this case any x from the domain will be a solution of the inequality because its left side is nonnegative whereas its right side is negative.

2: $x - 2 \geq 0$, or $x \geq 2$. In this case our inequality is equivalent to the following:

$$\frac{x^3 - 8}{x} > (x - 2)^2.$$

Since in this case $x > 0$, we get

$$x^3 + 8 > x(x - 2)^2,$$

that is,

$$x^3 + 8 > x^3 - 4x^2 + 4x$$
$$\Rightarrow 4x^2 - 4x + 8 > 0$$
$$\Rightarrow (2x - 1)^2 + 7 > 0,$$

which is true for any x from the domain.

Thus, we finally get the union of two intervals:

$$(-\infty, -2) \cup (0, +\infty). \qquad \square$$

3.21. Solve the following inequality:

$$x^5 - 5x^3 + 4x > 0.$$

Solution. As we discovered in Problem 2.1 (p. 19),

$$P(x) = x^5 - 5x^3 + 4x = (x - 2)(x - 1)x(x + 1)(x + 2).$$

Let us now draw the number line and plot the roots of $P(x)$:

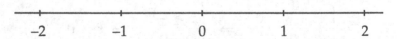

$P(x)$ is the product of five factors, all of which are positive when $x > 2$. As we move left along the number line, exactly one factor changes sign as we pass each point plotted.

This observation gives away the solution of the inequality; all we have to do is to draw a sine-like curve through all of the plotted points, which begins above the number line where $x > 2$:

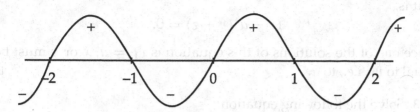

Since we are interested in finding x for which $P(x) > 0$, the answer is

$$(-2, -1) \cup (0, 1) \cup (2, \infty).$$ □

3.22. Prove that if (x, y, z) is a solution to the system of equations

$$\begin{cases} x + y + z = a \\ \dfrac{1}{x} + \dfrac{1}{y} + \dfrac{1}{z} = \dfrac{1}{a} \end{cases}$$

then at least one of the numbers x, y, z is equal to a.

Solution. Let (x, y, z) be a solution to the given system. Then, from the second equation,

$$\frac{1}{x} + \frac{1}{y} = \frac{1}{a} - \frac{1}{z},$$

that is,

$$\frac{x + y}{xy} = -\frac{a - z}{az}.$$

But from the first equation $x + y = a - z$; therefore,

$$xy(a - z) = -az(a - z).$$

Assume that $z \neq a$. Then, dividing by $(a - z)$, we get $xy = -az$. So we obtain the following two equalities:

$$x + y = a - z$$
$$xy = -az$$

By the converse of the Viète theorem, $\{x, y\}$ is the solution set of the following quadratic equation:

$$v^2 - (a - z)v - az = 0,$$

that is,

$$(v - a)(v + z) = 0.$$

Since one of the solutions of this equation is $v_1 = a$, x or y must be equal to v_1, i.e., to a.   □

3.23. Solve the following equation:

$$x^2 + \frac{1}{x^2} = 2^{1-y^2}.$$

Solution. As we know from Section 3.1, for any nonnegative a, b,

$$a + b \geq 2\sqrt{ab}.$$

Therefore,

$$x^2 + \frac{1}{x^2} \geq 2\sqrt{x^2 \frac{1}{x^2}} = 2.$$

So the minimum of the left side of the equation is 2. On the other hand, the maximum of the right side is 2. Therefore, the given equation is equivalent to the system

$$\begin{cases} x^2 + \dfrac{1}{x^2} = 2 \\ 2^{1-y^2} = 2. \end{cases}$$

By solving the equations of the system separately, we get $x = \pm 1$; $y = \pm 1$; so we have four solutions:

$$\begin{array}{cccc} x_1 = 1 & x_2 = 1 & x_3 = -1 & x_4 = -1 \\ y_1 = 1 & y_2 = -1 & y_3 = 1 & y_4 = -1 \end{array} \qquad □$$

Problems

3.24. Solve the following equation:

$$\frac{x^2}{3} + \frac{48}{x^2} = 10\left(\frac{x}{3} - \frac{4}{x}\right).$$

3.25. Solve the following system of equations:

$$4x^2 + 9y^2 = 10xy + 12$$
$$2x + 3y = 2xy.$$

3.26. Solve the following equation:

$$x^x = x.$$

3.27. Solve the following equation:

$$\sqrt{x^2 - 4x + 4} = -(x - 2).$$

3.28. Solve the following equation:

$$9 - x^2 = 2x\left(\sqrt{10 - x^2} - 1\right).$$

3.29. Solve the following equation:

$$\sqrt{x + 3 - 4\sqrt{x - 1}} + \sqrt{x + 8 - 6\sqrt{x - 1}} = 1.$$

3.30. Solve the following inequality:

$$\frac{x^2 - 7x + 12}{x^2 + x - 20} \geq 0.$$

3.31. Solve the following inequality:

$$\frac{(x + 1)(x + 3)^2(x + 5)^3}{(x + 2)^3(x + 4)^2(x + 6)} < 0.$$

3.32. (*Moscow State University, Entrance Examination, 1966*) Find all values of a such that the system of equations in x and y

$$\begin{cases} 2^{|x|} + |x| = y + x^2 + a \\ x^2 + y^2 = 1 \end{cases}$$

has exactly one solution.

4

Geometry

In this chapter we will use the following common notations: $|AB|$ for the length of a segment $\overline{AB}$; $m(\alpha)$ for the measure of an angle α; and $S(ABC)$ for the area of a triangle ABC.

4.1 Loci

We start with four familiar loci on the plane.

4.1. Given a point O and a positive real number r. The locus of all points P at the distance r from O is the circle of radius r with the center O (prove it!).

4.2. The locus of all points equidistant from two distinct points A and B is the perpendicular bisector of the segment (see Problem 1.16).

4.3. The locus of all points equidistant from two given intersecting lines is a pair of perpendicular lines that bisect all four angles between the given lines (prove it!).

Let F be a geometric figure and P a point. We say that F is *seen* from P at the angle α if α is the angle between two tangent rays drawn from P to F.

4.4. Given a segment $\overline{AB}$ and an angle α, $0 < \alpha < \pi$. The locus of all points P from which the segment $\overline{AB}$ is seen at the angle α (i.e., $APB = \alpha$) is a pair of arcs symmetric with respect to the line AB, excluding the arcs' endpoints A and B.

A. Soifer, *Mathematics as Problem Solving*, DOI: 10.1007/978-0-387-74647-0_4,

Solution. To begin, let us consider only the half-plane above the line $\overline{AB}$. We can draw an isosceles triangle AOB ($|AO| = |OB|$) with $\angle AOB = 2\alpha$ (Figure 4.1(a)), and the circle of radius $|OA|$ with the center in O.

For any point P on the arc AxB, the angle APB is equal to α, because it is measured by half of the arc AyB (check your geometry textbook for this theorem, but don't close your textbook yet!).

If we take any point P outside the circle (Figure 4.1(b)), then $\angle APB$ is less than α, because it is measured by half the difference between arcs AyB and CzD (textbook!).

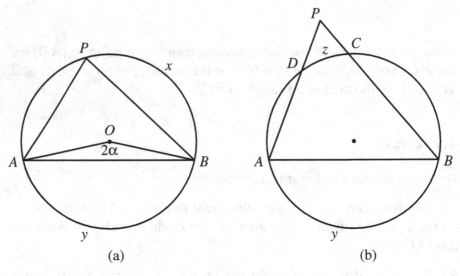

Fig. 4.1.

If we take any point P inside the circle (Figure 4.2(a)), then $\angle APB$ is greater than α, because it is measured by half the sum of arcs AyB and CxD. (Now you can close your textbook!)

Finally, "the story" in the half-plane below the line $\overline{AB}$ is symmetric to the one above AB; therefore, we get two symmetric arcs AxB and AyB, excluding the points A and B, of course (Figure 4.2(b)). □

4.5. Given a point O and a line L. Find the locus S of all second endpoints P of segments for which one endpoint lies on L and whose midpoint is at the point O.

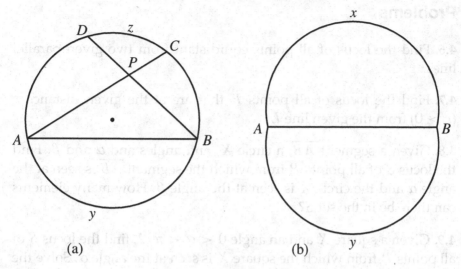

Fig. 4.2.

Solution. If a point P' lies on L, and O is the midpoint of $\overline{P'P}$, then P is the image of P' under rotation through the angle π about the point O. Therefore, the set S is the result of rotation of L through the angle π about O.

Of course, S is the line parallel to L the same distance from O as L (see Figure 4.3). □

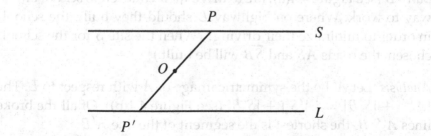

Fig. 4.3.

Problems

4.6. Find the locus of all points equidistant from two given parallel lines.

4.7. Find the locus of all points P that are at the given distance r $(r > 0)$ from the given line L.

4.8. Given a segment $\overline{AB}$, a circle X, and angles and α and β. Find the locus S of all points P from which the segment $\overline{AB}$ is seen at the angle α and the circle X is seen at the angle β. How many elements can there be in the set S?

4.9. Given a square X and an angle $0 < \alpha < \pi/2$, find the locus S of all points P from which the square X is seen at the angle α. Solve the same problem for $\pi/2 \le \alpha < \pi$.

4.10. Given segments $\overline{AB}$ and $\overline{XY}$ of lengths $|AB| = a$ and $|XY| = b$, $a \le b$. Find the locus of all points P such that $|PX|^2 - |PY|^2 = a^2$.

4.11. Prove that the three perpendicular bisectors of the sides of an arbitrary triangle intersect at one point.

4.2 Symmetry and Other Transformations

Symmetry

4.12. People living in the neighborhood A and working at the company B (see Figure 4.4(a)) are to drive their children to school on their way to work. Where on highway L should they build the school S in order to minimize their driving? (When the site S for the school is chosen, the roads $\overline{AS}$ and $\overline{SB}$ will be built.)

Analysis. Let A' be the symmetric image of A with respect to L. Then $|AS'| + |S'B| = |A'S'| + |S'B|$ (see Figure 4.4(b)). Of all the broken lines $A'S'B$, the shortest is the segment of the line $A'B$.

Solution. Draw the symmetric image A' of A with respect to L and the straight line $A'B$. The intersection S of $A'B$ and L is the site for the school. □

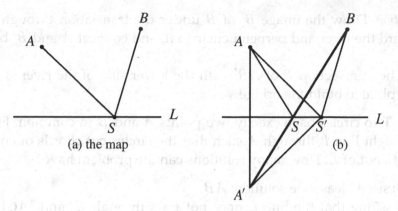

Fig. 4.4.

4.13. A river has straight parallel sides and cities A and B lie on opposite sides of the river (see Figure 4.5(a)). Where should we build a
bridge in order to minimize the traveling distance between A and B
(a bridge, of course, must be perpendicular to the sides of the river)?

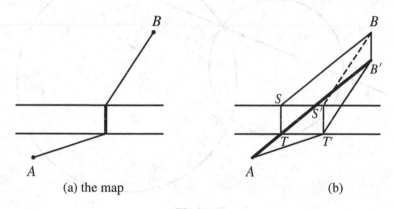

Fig. 4.5.

Analysis. Let B' be the result of the translation of B through the width
W of the river toward the river and perpendicular to it (see Figure 4.5(b)). Then

$$|AT'| + |T'S'| + |S'B| = (|AT'| + |T'B'|) + W.$$

Among all broken lines $AT'B'$, the shortest is the segment of the line
AB'.

Solution. Draw the image B' of B under the translation through W toward the river and perpendicular to it, and connect A and B' by a line.

The intersection T of AB' with the lower side of the river is the best place to build the bridge.   □

4.14. Two circles have exactly two points A and B in common. Find a straight line L through A such that the circles cut chords of equal length out of L. How many solutions can the problem have?

Analysis. At least one solution: AB.

Assume that the line L does not pass through B, and $|AC| = |AD|$ (see Figure 4.6).

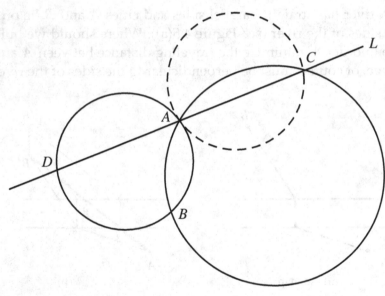

Fig. 4.6.

Since the points D and C are symmetric with respect to A, the symmetric image R_1' of the given circle R_1 through D must pass through C.

Solution. Draw the symmetric image R_1' of the given circle R_1. Every intersection C of R_1' with the other given circle R_2 determines the required line CA. The proof is obvious and essentially repeats the analysis.

Research. Since the point A is on R_2, and the circle R_1 has a point X inside R_2 (remember, R_1 and R_2 have exactly two distinct points in common!), the circle R_1' will have a point X' outside of R_2 (namely, the symmetric image X' of X with respect to A). Therefore, R_1' and R_2 have exactly one more point of intersection in addition to A, and thus the problem always has exactly two solutions (remember, the first solution was AB). □

I have no doubts that my readers have a good knowledge of some frequently used types of transformations: translation, rotation, central symmetry, and line symmetry. I would like to say a few words here about another very important type of transformation: homothety.

Homothety

Given a point O and a nonzero number k. A *homothetic transformation*, or *homothety*, H with the *center of homothety* O and *coefficient of homothety k* is the transformation that maps every point P into the point $P_1 = H(P)$ on the straight line OP such that

$$\frac{|\overrightarrow{OP_1}|}{|\overrightarrow{OP}|} = k.$$

Please note that unlike the case with the length of a segment, the direction of vectors $\overrightarrow{OP_1}$ and $\overrightarrow{OP}$ is taken into account. The measure $|\overrightarrow{a}|$ of a vector $\overrightarrow{a}$ is equal to the length $|a|$ of the corresponding

$$O \qquad\qquad P \qquad P_1$$

Fig. 4.7. Homothety with $k = 3/2$.

segment $\overline{a}$ if the direction of $\overrightarrow{a}$ is the same as that of a unit vector, and is equal to $-|a|$ otherwise. A negative coefficient of homothety k would imply that the points P_1 and P are on opposite sides of the center of homothety O.

The *homothetic image* $F_1 = H(F)$ of a geometric figure F under the homothety H is the geometric figure consisting of homothetic images $P_1 = H(P)$ of all points P that make up the given figure F.

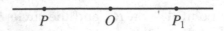

Fig. 4.8. Homothety with $k = -1$.

Please note (and prove!) that:

(a) The homothetic image $I_1 = H(I)$ of a segment I is the segment I_1 parallel to I such that $|I_1|/|I| = k$, where k is the coefficient of homothety.
(b) Two homothetic polygons $H(P)$ and P are similar.
(c) Any two circles K_1, K_2 are homothetic (i.e., there exists a homothety H such that $K_2 = H(K_1)$).
(d) Symmetry with respect to a point is a particular case of homothety with the coefficient of homothety $k = -1$.

4.15. Inscribe a square in a given acute triangle.

Analysis. Let the square $MNPQ$ be inscribed in the triangle ABC. Two points of the square, say P, Q, must lie on the same side of the triangle, say $\overline{AC}$, with two others, one per side of the triangle (see Figure 4.9).

   If we waive the requirement that the vertex N of the square be on $\overline{BC}$, we will get many (in fact, infinitely many) squares satisfying all other conditions (i.e., $\overline{PQ}$ lies on $\overline{AC}$, and M on $\overline{AB}$). Let us take one of them, $M_1 N_1 P_1 Q_1$. Then the squares $M_1 N_1 P_1 Q_1$ and $MNPQ$ are homothetic!

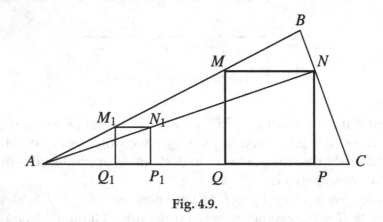

Fig. 4.9.

Construction. (1) Construct a square $M_1 N_1 P_1 Q_1$ with $\overline{P_1 Q_1}$ on $\overline{AB}$ and M_1 on $\overline{AB}$: pick a point M_1 on $\overline{AB}$, draw $\overline{M_1 Q_1} \perp \overline{AC}$, mark P_1 such that $|Q_1 P_1| = |M_1 Q_1|$, draw the perpendicular to $\overline{AC}$ at P_1, then draw the perpendicular to $\overline{M_1 Q_1}$ at M_1, with their intersection determining N_1.

(2) Draw the straight line through A and N_1. Denote the intersection of AN_1 and BC by N.

(3) The homothety H with the center in A and the coefficient $k = |AN|/|AN_1|$ concludes the construction:

$$H(M_1 N_1 P_1 Q_1) = MNPQ.$$

Proof. The homothetic image of a square is a square. $N = H(N_1)$ was chosen to lie on $\overline{BC}$. $M = H(M_1)$ lies on the line AM_1. Similarly $P = H(P_1)$ and $Q = H(Q_1)$ lie on AC. Finally, since $NM \parallel AC$ and N is an inside point of $\overline{BC}$, M is an inside point of the segment $\overline{AB}$. Similarly, you can show that P and Q are inside points of the segment $\overline{AC}$.

Research. The existence and uniqueness of the point N in our construction gives us exactly one solution under the assumption that two vertices of the square lie on $\overline{AC}$. We can construct another solution with two vertices of the square on $\overline{AB}$, and a third one with two vertices of the square on $\overline{BC}$.

Thus the problem always has exactly three solutions. □

Problems

4.16. Where would you build two bridges over the two sleeves of a river with parallel straight sides (see Figure 4.10) to minimize the length of the path between the cities A and B? (Bridges have to be perpendicular to the sides of the river.)

4.17. Given three parallel lines L_1, L_2, L_3. Find an equilateral triangle ABC with A on L_1, B on L_2, and C on L_3.

4.18. Given three concentric circles K_1, K_2, K_3. Find an equilateral triangle ABC with A on K_1, B on K_2, and C on K_3.

4.19. Given a point A and a circle K with center O. Connect A with an arbitrary point B of the circle K. Find the locus of a point P of the intersection of AB with the bisector of the angle AOB.

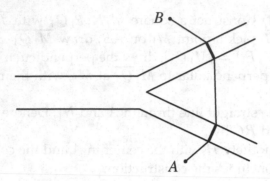

Fig. 4.10. The Map.

4.3 Proofs in Geometry

4.20. Prove that if a and b are two sides of a triangle and m_c is the median drawn to the third side, then

$$|m_c| \leq \frac{|a| + |b|}{2}.$$

Solution. Let $|BO| = |OA|$ (Figure 4.11). We extend CO and mark D so that $|CO| = |OD|$. Then the quadrilateral $ACBD$ is a parallelo-

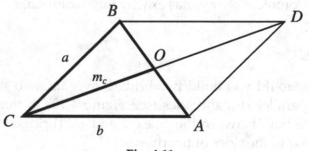

Fig. 4.11.

gram. Therefore, $|BD| = |CA| = |b|$. Finally, in the triangle CBD, we have

$$|CD| < |CB| + |BD|,$$

i.e.,

$$2|m_c| < |a| + |b|.$$

Therefore,

$$m_c < \frac{|a| + |b|}{2}. \qquad \Box$$

4.21. A circle is inscribed in a triangle ABC. $\overline{MN}$ is the diameter perpendicular to the base $\overline{AC}$. Let L be the intersection of BM with AC (see Figure 4.12). Prove that $|AN| = |LC|$.

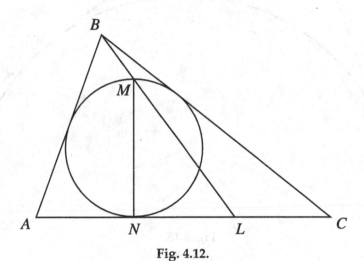

Fig. 4.12.

We will show here two solutions: a geometric solution and an algebraic one.

Solution I

Draw the tangent line EF to the circle at the point M. $EF \parallel AC$. Consider the homothetic transformation T with center B mapping M into L. T will map $\overline{EF}$ onto $\overline{AC}$ and the given circle into the circle tangent to BA, BC, and AC outside of the triangle ABC (see Figure 4.13).

Denote the tangent points of the circles and BA and BC by X, Y, Z, and W, respectively.

Since the lengths of the segments of two tangent lines drawn from a point to a circle are equal, we get:

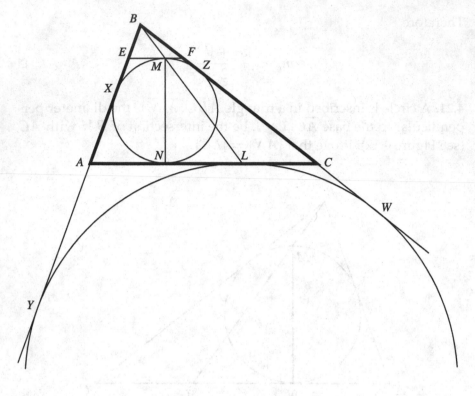

Fig. 4.13.

$$|BY| = |BW|$$
$$-|BX| = |BZ|$$
$$\overline{}$$
$$|XY| = |ZW|.$$

But $|AX| = |AN|$ and $|AY| = |AL| = |AN| + |NL|$. Therefore,

$$|XY| = 2|AN| + |NL|.$$

Similarly $|CZ| = |CN| = |NL| + |LC|$ and $|CW| = |CL|$. Therefore,

$$|ZW| = 2|LC| + |NL|.$$

Finally, we have

$$2|AN| + |NL| = 2|LC| + |NL|,$$

i.e.,

$$|AN| = |LC|.$$ □

Solution II

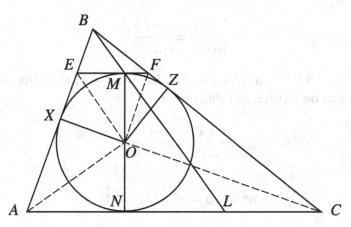

Fig. 4.14.

Let EF be the tangent line through M. Then $EF \parallel AC$ (see Figure 4.14). Let X and Z be the tangent points of the circle and BA and BC, respectively. We connect A, C, F, and E with the center O of the circle.

Let $m(\alpha)$ denote the measure of angle α. Since

$$m(\angle OAE) + m(\angle OEA) = \tfrac{1}{2}m(\angle CAE) + \tfrac{1}{2}m(\angle FEA) = \tfrac{1}{2}\pi,$$

AOE is a right triangle; similarly, FOC is a right triangle. Therefore,

$$r^2 = |OX|^2 = |EX| \cdot |XA| = |EM| \cdot |AN|$$
$$r^2 = |OZ|^2 = |FZ| \cdot |ZC| = |FM| \cdot |CN|.$$

Thus,

$$|EM| \cdot |AN| = |FM| \cdot |CN|,$$

i.e.,

$$\frac{|EM|}{|MF|} = \frac{|CN|}{|AN|}.$$

On the other hand, due to the homothetic transformation with center B mapping M into L, we have

$$\frac{|EM|}{|MF|} = \frac{|AL|}{|LC|}.$$

Combining the two equalities above, we get

$$\frac{|CN|}{|AN|} = \frac{|AL|}{|LC|}.$$

If we let $|AN| = a$, $|NL| = b$, and $|LC| = c$, then the previous equality can be written as follows:

$$\frac{b+c}{a} = \frac{a+b}{c},$$

i.e.,

$$a^2 + ab - c^2 - bc = 0,$$

or

$$(a - c)(a + b + c) = 0.$$

Therefore,

$$a = c.$$

Thus $|AN| = |LC|$. □

Problems

4.22. Prove that in any triangle, a bisector lies between the altitude and the median drawn from the same vertex of the triangle.

4.23. Prove that diagonals d_1, d_2 and sides a_1, a_2, a_3, a_4 of a parallelogram satisfy the following equality:

$$|d_1|^2 + |d_2|^2 = |a_1|^2 + |a_2|^2 + |a_3|^2 + |a_4|^2.$$

4.24. An equilateral triangle ABC is inscribed in a circle. Prove that no matter where we chose a point P on the arc AB, $|PC| = |PA| + |PB|$.

4.25. Prove that the medians of any triangle are themselves sides of a triangle.

4.26. (Ceva Theorem) The points M, N, K are on opposite sides of the triangle ABC (see Figure 4.15). If AM, BN, and CK intersect at one point, then

$$\frac{|AK|}{|KB|} \cdot \frac{|BM|}{|MC|} \cdot \frac{|CN|}{|NA|} = 1,$$

and the converse.

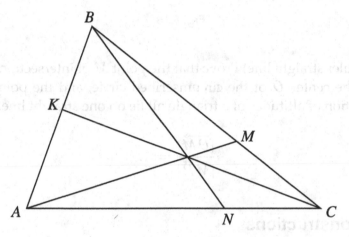

Fig. 4.15.

4.27. Prove that the line through the intersection of the extensions of the nonparallel sides of a given trapezoid and the intersection of its diagonals bisects both parallel bases of the trapezoid.

4.28. Prove that if the line through the intersection of the extensions of the opposite sides AB and CD of the given convex quadrilateral $ABCD$ and the intersection O of its diagonals AC and BD bisects its base AD or its base BC, then the given quadrilateral is a trapezoid, namely $BC \parallel AD$ (see Figure 4.16).

4.29. Prove that the altitudes of an arbitrary triangle intersect at one point.

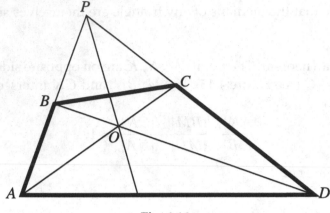

Fig. 4.16.

4.30. (Euler straight line) Prove that the point M of intersection of medians, the center O of the circumscribed circle, and the point H of intersection of altitudes of a triangle all lie on one straight line. Moreover,

$$\frac{|\overrightarrow{OM}|}{|\overrightarrow{MH}|} = \tfrac{1}{2}.$$

4.4 Constructions

In this section, we will show how loci and transformations work in geometrical constructions. We will also look into constructions with limited means (i.e., compass and ruler).

4.31. Given an angle $ABC, 0 < m(\angle BAC) < \pi$, and a point O inside it. Construct a line L such that O is the midpoint of the segment cut out of the line L by the sides BA and BC of the given angle.

Analysis. Assume that MN is the required line, where M is on BA and N is on BC (see Figure 4.17(a)).

Let us forget for a moment that N lies on BC (Figure 4.17(b)). Then all we know about N is that it is the second endpoint of a segment with one endpoint M on the given line BA and with midpoint at O. Therefore (see Problem 4.5), N lies on the line S, which is the result of rotation of BA about O through the angle π.

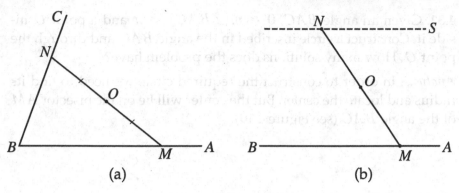

Fig. 4.17.

Now the intersection of S and BC will give us the point N.

Thus all we need to know now is how to draw S. A line is determined by two points, so if we pick two points M_1, M_2 on the line BA and rotate each of them about O through the angle π, we get two points M_1', M_2' on the line S.

Construction. (1) Pick two distinct points M_1, M_2 on BA (see Figure 4.18).

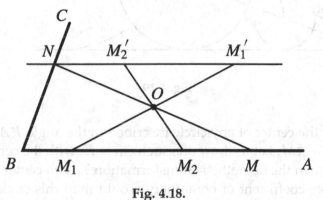

Fig. 4.18.

(2) Draw the line through M_1 and O, and mark a point M_1' on it such that $|M_1 O| = |O M_1'|$.

(3) Draw the line through M_2 and O, and mark a point M_2' on it such that $|M_2 O| = |O M_2'|$.

(4) Draw the line through M_1' and M_2'. Denote its intersection with BC by N.

(5) The required line is NO.   □

4.32. Given an angle BAC, $0 < m(\angle BAC) < \pi$ and a point O inside it. Construct a circle inscribed in the angle BAC and through the point O. How many solutions does the problem have?

Analysis. In order to construct the required circle we need to find its radius and locate the center. But the center will lie on the bisector AM of the angle BAC (see Figure 4.19).

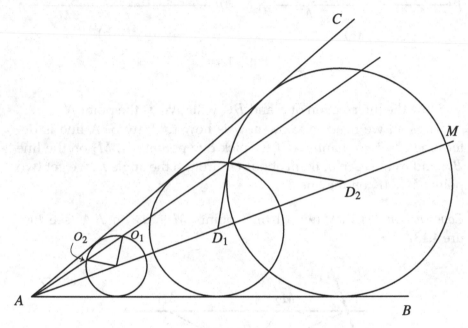

Fig. 4.19.

In fact the center of any circle inscribed in the angle BAC lies on the bisector AM. Let us draw one such circle K, with the center at the point D. Then the homothetic transformation T with center A and an appropriate coefficient of homothety would map this circle into the required circle. But under the homothety T, the point O would be the image of a point that is both on the circle K and also on the line AO. We can find such a point; in fact, there are two of them, O_1 and O_2, the intersections of AO and the circle K. Thus we have two solutions: the image K_1 of the circle K under the homothetic transformation T_1 with center A and the coefficient determined by $T_1(O_1) = 0$, and the image K_2 of the circle K under the homothetic transformation T_2 with center A and the coefficient determined by $T_2(O_2) = O$.

Construction. We present here an outline of construction and recommend the reader completely describe the construction and provide all the proofs.

(1) Construct the bisector AD of the angle BAC.

(2) Inscribe a circle K in the angle BAC. Denote its center by D.

(3) Draw the straight line AO. Denote its intersections with the circle K by O_1 and O_2.

(4) Draw the line DO_1.

(5) Construct the line through O parallel to DO_1. Denote its intersection with AM by D_1.

(6) Draw the required circle K_1 with center D_1 and radius $|D_1O|$.

In order to find the second solution K_2, replace $O1$ by $O2$ in (4), (5), and (6). □

4.33. Given segments $\bar{a}$ and $\bar{b}$. Construct the segment of length $\sqrt[4]{|a|^4 + |b|^4}$ with compass and ruler (note that the unit length is not given!).

Analysis.

$$\sqrt[4]{|a|^4 + |b|^4} = \sqrt[4]{|a|^2 \left(|a|^2 + \frac{|b|^4}{|a|^2}\right)} = \sqrt{|a| \sqrt{|a|^2 + \left(\frac{|b|^4}{|a|}\right)^2}}$$

This reveals the plan of our construction!

Construction. (1) Construct the segment of length $|b|^2/|a|$. One way to do this is to construct a right triangle with height b drawn from the right angle to the hypotenuse and the projection of one side onto the hypotenuse equal to $\bar{a}$ (see Figure 4.20).

So, we draw

(a) $\overline{AB}$ such that $|AB| = |a|$;
(b) $CB \perp AB$;
(c) the point D such that $|DB| = |b|$;
(d) the line through A and D;
(e) $DE \perp AD$.
(f) Finally, denote the intersection of DE and AB by F.

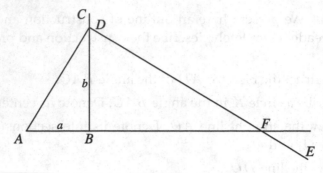

Fig. 4.20.

Then the length $\overline{BF}$ is equal to $|b|^2/|a|$ (prove it!)

    (2) Construct the segment $\overline{c}$ of length

$$\sqrt{|a|^2 + \left(\frac{|b|^4}{|a|}\right)^2}.$$

The hypotenuse of the right triangle with legs of lengths $|a|$ (given) and $|b|^2/|a|$ (constructed above) has the required length.

    (3) Construct the segment of length $\sqrt{|a|\,|c|}$ where $\overline{c}$ is constructed in (2). One way to do this is to construct a right triangle with projections of legs onto the hypotenuse equal to $\overline{a}$ and $\overline{c}$ and take the altitude drawn from the right angle to the hypotenuse (see Figure 4.21).

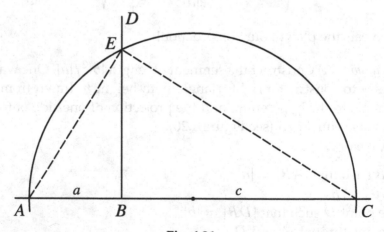

Fig. 4.21.

(a) Mark $|AB| = |a|$ and $|BC| = |c|$.

(b) Draw a circle with diameter $\overline{AC}$.

(c) Draw $BD \perp AC$.

(d) Denote by E the intersection of BD with the circle.

Then the length of $\overline{BE}$ of is equal to $\sqrt{|a|\,|c|}$ (prove it!), and

$$\sqrt{|a|\,|c|} = \sqrt{|a|\sqrt{|a|^2 + \left(\frac{|b|^4}{|a|}\right)^2}} = \sqrt{|a|^4 + |b|^4}. \qquad \square$$

4.34. Given a circle K, its diameter $\overline{AB}$, and a point O outside of K and not on the line AB. With a straight edge alone, construct the perpendicular to AB through O. (Note that the center of the circle is not given!)

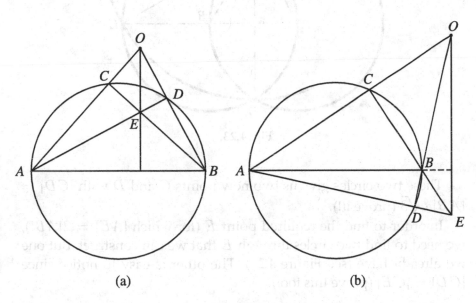

(a)   (b)

Fig. 4.22.

Analysis. If we draw lines OA and OB and denote their intersections with the circle K by C and D, respectively, then $m(\angle ACB) = m(\angle ADB) = \pi/2$ (see Figure 4.22 (a) and (b), representing two possible cases: either both $\angle OAB$ and $\angle OBA$ are acute, or one of them is obtuse. The third case, when $\angle OAB$ or $\angle OBA$ is equal to $\pi/2$, is trivial).

If we denote the intersection of AD and BC by E, the line OE will be the required perpendicular to AB because three altitudes of a triangle intersect at one point! The construction is clear from the analysis and left for the reader.   □

4.35. With a compass alone, double the given segment $\overline{AB}$.

Analysis. All we can do to begin is draw two circles of radius $\overline{AB}$ with centers A and B (see Figure 4.23).

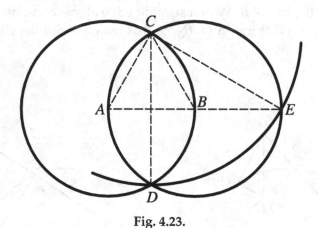

Fig. 4.23.

These two circles give us two new points C and D with $|CD| = |AB| \cdot \sqrt{3}$ (prove it!).

In order to find the required point E (for which $|AE| = 2|AB|$), we need to find two circles through E that we can construct. But one we already have (see Figure 4.23). The other is easy to notice since $|CD| = |CE|$ (prove this too!).

Construction. (1) Draw the circles K_1 and K_2 of radius $|AB|$ with centers at A and B, respectively. Denote their intersections by C and D.

(2) Draw the circle K_3 of radius $|CD|$ with center at C. Its intersection E with K_2 gives us the required segment $\overline{AE}$.

The proof has essentially been done above by the reader (I hope).

□

There is a second solution based on the observation that the side of a regular hexagon is equal to the radius of the circumscribed circle. The implementation of this idea is left to the reader.

Problems

4.36. Given a line L and two points A, B on one side of it. Construct a circle through A and B tangent to L. How many solutions does the problem have?

4.37. Given segments $\bar{a}$ and $\bar{b}$, $|a| \leq |b|$. With compass and straight edge, construct the segment of length $\sqrt[4]{|a|^4 - |b|^4}$.

4.38. (Moscow Mathematical Olympiad, 1966) Divide the given segment $\bar{a}$ into six segments of equal length by drawing only eight lines with compass and straight edge.

4.39. Given two parallel lines L_1 and L_2 and a point P. With a straight edge alone construct the line L parallel to L_1 and L_2 through P. (Hint: first solve Problem 4.28, then use it!)

4.40. With compass alone divide a given segment in half. (Hint: use Problem 4.35 discussed in this section.)

4.5 Computations in Geometry

4.41. Given the lengths a, b, and c of the sides of a triangle T. Compute

$$(h_a + h_b + h_c)\left(\frac{1}{h_a} + \frac{1}{h_b} + \frac{1}{h_c}\right),$$

where h_a, h_b, h_c are the lengths of the corresponding altitudes of T.

Solution. If S denotes the area of the triangle T, then $2S = ah_a = bh_b = ch_c$ and we get

$$(h_a + h_b + h_c)\left(\frac{1}{h_a} + \frac{1}{h_b} + \frac{1}{h_c}\right)$$

$$= \left(\frac{2S}{a} + \frac{2S}{b} + \frac{2S}{c}\right)\left(\frac{a}{2S} + \frac{b}{2S} + \frac{c}{2S}\right)$$

$$= (a + b + c)\left(\frac{1}{a} + \frac{1}{b} + \frac{1}{c}\right). \qquad \square$$

4.42. Let E, F, and G be points on the sides $\overline{AB}$, $\overline{BC}$, and $\overline{CA}$ of the triangle ABC such that

$$\frac{|AB|}{|EB|} = \frac{|BF|}{|FG|} = \frac{|CG|}{|GA|} = k,$$

where $0 < k < 1$. Find the ratio of the area of the triangle KML, created by the intersections of the lines AF, BG, CE to the area of the triangle ABC (see Figure 4.24).

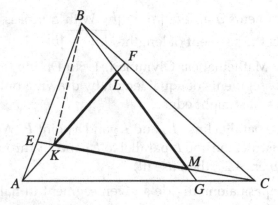

Fig. 4.24.

Solution. Denote the area of a triangle XYZ by $S(XYZ)$. Then, by using the fact that the ratio of the areas of two triangles with the same heights (or the same bases) is equal to the ratio of their bases (or the ratio of their heights, respectively), and that $S(AEC) = S(AEK) + S(ACK)$, we get:

$$S(ACK) = \frac{1}{k}S(ABK) = \frac{k+1}{k^2}S(AEK)$$

$$= \frac{k+1}{k^2+k+1}S(AEC) = \frac{k}{k^2+k+1}S(ABC).$$

Similarly,

$$S(BLA) = S(CMB) = \frac{k}{k^2+k+1}S(ABC).$$

Therefore,

$$\frac{S(KLM)}{S(ABC)} = 1 - \frac{3k}{k^2 + k + 1} = \frac{(1-k)^2}{k^2 + k + 1} = \frac{(1-k)^3}{1 - k^3}. \qquad \square$$

Not only are loci, transformations, and constructions helpful in solving computational problems. Sometimes computational problems provide the only clues for solving other types of problems.

4.43. (*A. Soifer, 1971*) Partition an arbitrary triangle using six straight cuts into parts from which one can assemble seven congruent triangles.

Analysis. In the previous problem we had

$$\frac{S(KLM)}{S(ABC)} = \frac{(1-k)^3}{1 - k^3}.$$

It is easy to see that for $k = \frac{1}{2}$, $\frac{(1-k)^3}{1-k^3} = \frac{1}{7}$. So if we partition the sides of a triangle in the ratio $1 : 2$,

$$\frac{|AE|}{|EB|} = \frac{|BF|}{|FC|} = \frac{|CG|}{|GA|} = \frac{1}{2},$$

and then make cuts AF, BG, CE, we will have one part, the triangle KLM, of exactly the right size: $S(KLM) = \frac{1}{7}S(ABC)$.

We also notice that if $k = \frac{1}{2}$, each of the three segments $\overline{AF}, \overline{BG},$ and $\overline{CE}$ is split by the two others in the ratio $3 : 3 : 1$ (starting with the angles of the triangle ABC).

Indeed, the chain of the area equalities of the previous problem with $k = \frac{1}{2}$ gives us $S(AEK) = \frac{1}{7}S(AEC)$. Therefore,

$$|EK| = \frac{1}{7}|EC|. \qquad (*)$$

Similarly, $|GM| = \frac{1}{7}|GB|$ and $S(ABK) = S(ACM)$. But the same chain of equalities shows that $S(ABK) = \frac{1}{2}S(ACK)$; therefore, $S(ACM) = \frac{1}{2}S(ACK)$, i.e.,

$$|CM| = \frac{1}{2}|CK|. \qquad (**)$$

The two equalities (∗) and (∗∗) above prove that $|CM| : |MK| : |KE| = 3 : 3 : 1$.

Now it is not hard to find the remaining three cuts; we draw them through the points M, K, L parallel to AF, BG, CE respectively. They complete the partition of the sides of the given triangle in the ratios $2 : 1 : 1 : 2$.

Construction. Behold! (See Figure 4.25.)

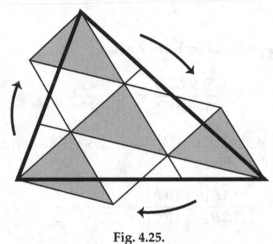

Fig. 4.25.

The proof is essentially contained in our detailed analysis.   □

Problems

4.44. Prove that

$$h_a + h_b + h_c = \frac{ab + bc + ca}{2R},$$

where a, b, c, h_a, h_b, h_c, and R are the lengths of the sides, altitudes, and radius of the circumscribed circle of a given triangle, respectively.

4.45. Prove that

$$\frac{1}{h_a} + \frac{1}{h_b} + \frac{1}{h_c} = \frac{1}{r},$$

where h_a, h_b, h_c, and r are the lengths of the altitudes and radius of the inscribed circle of a given triangle, respectively.

4.46. Given three sides a, b, c of a triangle. Find its medians.

4.47. The area of a triangle is equal to S. Find the area of the triangle whose sides are the medians of the given triangle (see Problem 4.25).

4.48. The area of the equilateral triangle constructed on the hypotenuse of a right triangle is twice the area of the right triangle. Find the ratio of the legs of the right triangle.

4.49. Given the area S and the radius r of the circumscribed circle of a triangle. Find the product of the lengths of its three sides.

4.50. Given a right triangle ABC with right angle A. The altitude $\overline{AK}$ is drawn to the hypotenuse; from K the perpendiculars $\overline{KP}$ and $\overline{KT}$ are drawn to the legs $\overline{AB}$ and $\overline{AC}$, respectively. If $|BP| = m$ and $|CT| = n$, find the length of the hypotenuse $\overline{BC}$.

4.6 Maximum and Minimum in Geometry

4.51. Of all triangles with a given base and area, find the one with the minimal perimeter.

Analysis. If the area S and the base $|BC| = a$ of a triangle ABC are given, we can figure out the length of the altitude h_a from the equation $S = \frac{1}{2}ah_a$.

But the locus of all points A that are the given distance h_a from the given line BC in the half-plane above BC is the line L parallel to BC. L is the distance h_a from BC (Problem 4.7).

Since the perimeter $P = |BA| + |AC| + |BC|$ and $|BC| = a$, the minimal perimeter is attained exactly when $|BA| + |AC|$ is minimal.

Now we can reformulate the problem as follows: for given points B, C and a line L, find the point A on L that minimizes $|BA| + |AC|$ (see Figure 4.26).

But we have already solved this problem in Section 4.2 (Problem 4.12)!

In the particular case when $BC \parallel L$ (which we have here), the resulting triangle BAC is isosceles (prove it!).

The construction is left for the reader. The proof is essentially contained in the analysis above.   □

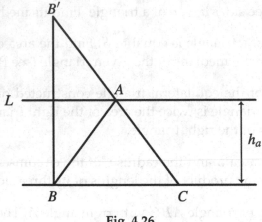

Fig. 4.26.

4.52. Given an angle $ABC, 0 < m(\angle ABC) < \pi$ and a point O inside it. Find the line through O that cuts the triangle of the minimal area out of the angle ABC.

Analysis. The setting of this problem might remind you of Problem 4.31. In that problem we constructed the line L such that the point O was the midpoint of the segment $\overline{MN}$ cut out of L by the sides BA and BC of the given angle. Let us compare $\overline{MN}$ to any other line through O (see Figures 4.27(a) and (b)). Let $ND \parallel BC$ in Fig-

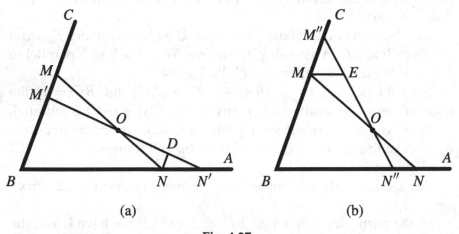

(a)   (b)

Fig. 4.27.

ure 4.27(a), and $ME \parallel BA$ in Figure 4.27(b).

Without any difficulties you can prove now (do!) that $S(MBN) <$ $S(M'BN')$ and $S(MBN) < S(M''BN'')$, where $S(XYZ)$ denotes the area of a triangle XYZ.

The construction is already done in Problem 4.31! The proof is contained in the analysis.   □

4.53. In the setting of the previous problem, find the line through O that cuts out of the angle ABC the triangle of minimal perimeter.

Analysis. (a) For the time being let us forget about the point O and inscribe a circle K in angle ABC. Denote the tangent points of K and BA and BC by M and N, respectively (see Figure 4.28).

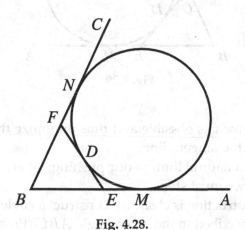

Fig. 4.28.

Let us now draw the tangent line EF through a point D of the circle K that is inside the triangle MBN. Since the lengths of the segments of two tangent lines drawn from a point to a circle are equal, we get $|BN| = |BM|, |EM| = |ED|$, and $|FD| = |FN|$.

Now we can figure out the perimeter $P(BEF)$ of the triangle BEF in terms of the length of the segment $\overline{BM}$ of the tangent line:

$$P(BEF) = (|BF| + |FD|) + (|BE| + |ED|)$$
$$= (|BF| + |FN|) + (|BE| + |EM|)$$
$$= |BN| + |BM| = 2|BM|.$$

(b) Let us put back the point O, draw a line through it intersecting the sides BA and BC of the given angle in the points E and F, and inscribe the circle K in the angle ABC tangent to EF outside of the triangle EBF (see Figure 4.29).

From part (a) we know that $P(EBF) = 2|BM|$. Therefore, in order to minimize $P(EBF)$ we need to "push" the circle K into the

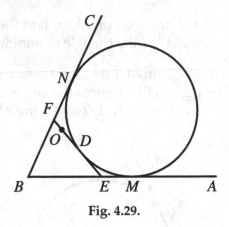

Fig. 4.29.

angle ABC as much as possible and thus minimize the length of the segment $\overline{BM}$ of the tangent line.

But we have a natural limit to our pushing: when the point O lies on the circle K, we must stop!

Now the construction is clear: we construct a circle K through the given point O inscribed in the given angle ABC (Problem 4.32!), and draw the tangent line to K at the point O.

It is important to choose the larger of the two circles constructed in Problem 4.32.

The construction is already done in Problem 4.32! The proof is essentially contained in our analysis above.   □

Problems

4.54. Of all triangles with two given sides, find the one of maximal area.

4.55. Of all triangles with a given side and opposite angle, find the one of maximal area.

4.56. Of all rectangles of a given area, find the one of minimal perimeter.

4.57. Of all rectangles inscribed in a given circle, find the one of the maximal area.

4.58. Given three points A, B, P on the plane. Find the line L through P that maximizes the sum of distances from A and B to L.

4.59. Given points A, B, P on the plane. Find the line through P that minimizes the sum of distances from A and B to L.

4.60. Of all triangles inscribed in a given circle, find the one of maximal area.

4.61. Given a positive integer n and a circle. Of all inscribed convex n-gons, find the one of maximal area.

4.62. Of all triangles inscribed in a given square, find the one of maximal area.

5

Combinatorial Problems

5.1 Combinatorics of Existence

Clearly, all the problems we discussed in Section 1.4 (Pigeonhole Principle) delivered existence results, and thus belong here, too.

There are, however, many other striking ideas used in combinatorial problems on existence. We will introduce a few of them here.

5.1. (*Second Annual Colorado Springs Mathematical Olympiad, 1985*) Each of the 49 entries of a 7×7 table is filled with an integer between 1 and 7 so that each column contains all of the integers 1, 2, 3, 4, 5, 6, 7, and the table is symmetric with respect to its diagonal D going from the upper left corner to the lower right corner. Prove that this diagonal D has all of the integers 1, 2, 3, 4, 5, 6, 7 on it.

Solution. The total number of entries 1 in the table is odd (one per column). For every entry 1 off the diagonal D, the table contains 1 in the square symmetric to the first one with respect to D. Therefore, the number of entries 1 off the diagonal D in the table is even. Thus D contains at least one entry 1.

Similarly, D contains at least one 2, at least one 3, ..., and at least one 7. ☐

5.2. (*After F.P. Ramsey*) Prove that in any party of six people there are three mutual acquaintances or three mutual non-acquaintances.

Solution. (a) It is convenient to record the information about acquaintances of a group of six people in two diagrams, D_1 and D_2. In both

A. Soifer, *Mathematics as Problem Solving*, DOI: 10.1007/978-0-387-74647-0_5,
© Alexander Soifer 2009

diagrams we represent each person by a vertex of a regular hexagon. Two vertices of D_1 are connected by a segment if and only if they correspond to two people who are acquainted. Two vertices of D_2 are connected by a segment if and only if they correspond to two people who are not acquainted.

Note that any two vertices are connected in exactly one of the two diagrams D_1 or D_2.

The problem is now equivalent to proving that at least one of the two diagrams D_1 or D_2 contains a triangle!

(b) Let us fix the same vertex A in both diagrams D_1 and D_2. A is connected with each of the five other vertices either in D_1 or in D_2; therefore, it must be connected with at least three of the five vertices in D_1 or in D_2 (Pigeonhole Principle with five pigeons and two holes!).

Due to the symmetry of the problem, we can assume without loss of generality that A is connected to the vertices B_1, B_2, and B_3 in D_1.

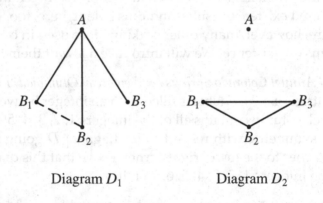

Diagram D_1 Diagram D_2

If any two of the vertices B_1, B_2, B_3 are connected in D_1, then these two vertices and A are the vertices of a triangle contained in D_1. If no two of the vertices B_1, B_2, B_3 are connected in D_1, then the triangle $B_1 B_2 B_3$ is contained in the diagram D_2.   □

It is interesting to note that six is the smallest number of people in the company that guarantees the above result. A party of five people might have neither three mutual acquaintances nor three mutual non-acquaintances. This can be proven by the following two diagrams, neither of which contains a triangle (please see "Diagram of acquaintances" and "Diagram of non-acquaintances" below).

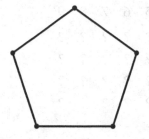

Diagram of
acquaintances

Diagram of
non-acquaintances

We will now discuss the same problem (please see Problem 1.23 in Section 1.4) once again.

5.3. Forty-one rooks are placed on a 10×10 chessboard. Prove that you can choose five of them that do not attack each other.

Solution. This solution was found in 1984 by George Berry, a software engineer at Digital Equipment Corporation. (No one else has found this solution independently ever since.) George's use of the linear algebra concept of matrix diagonalization is strikingly beautiful. Enjoy!

Suppose that you could choose at most four rooks that do not attack each other. Select any four such rooks. We will call these the key rooks.

Now consider an operation on the chessboard that exchanges two rows or two columns of the board. This exchange does not affect the attacks of any of the rooks.

We can use this operation to "normalize" the chessboard, that is, to put the four key rooks into locations $(0, 0)$, $(1, 1)$, $(2, 2)$, and $(3, 3)$, where the first coordinate determines the row and the second coordinate determines the column of a rook.

The normalization procedure can be done as follows: locate any key rook and exchange its row with row 0 and its column with column 0; do likewise for the other three key rooks (moving them into $(1, 1)$, etc.). At the end of normalization the chessboard looks like this:

	0	1	2	3	4	5	6	7	8	9
0	K	?	?	?	?	?	?	?	?	?
1	?	K	?	?	?	?	?	?	?	?
2	?	?	K	?	?	?	?	?	?	?
3	?	?	?	K	?	?	?	?	?	?
4	?	?	?	?	-	-	-	-	-	-
5	?	?	?	?	-	-	-	-	-	-
6	?	?	?	?	-	-	-	-	-	-
7	?	?	?	?	-	-	-	-	-	-
8	?	?	?	?	-	-	-	-	-	-
9	?	?	?	?	-	-	-	-	-	-

- = there is no rook
K = occupied by key rook
? = do not know whether there is a rook there or not.

Consider the square of the chessboard with coordinates (n, m) or (m, n) such that $0 \leq n \leq 3$ and $4 \leq m \leq 9$. There are 48 such squares. We will call these the outside squares of the board.

Select a pair of outside squares with coordinates (n, m) and (m, n). Is it possible for both of these squares to contain a rook? No, for if both squares had rooks in them then we could select a new set of key rooks by taking the rooks on those two squares along with three of the four key rooks (omitting the one at (n, n)). This would form a new set of *five* key rooks; we eliminated the only one of the original key rooks that attacked (n, m) and (m, n), and obviously (n, m) does not attack (m, n). However, a set of *five* key rooks is impossible by the original assumption that the largest number of key rooks was four.

Therefore, only one of the pair of outside squares (n, m) and (m, n) is occupied. That means that there are at most 24 rooks on the outside squares (half the squares are occupied). When added to the 16 rooks (the most possible on the inside squares), we have a total of 40 rooks accounted for.

That is one less than the 41 rooks given in the problem, so the assumption that there are at most four key rooks must be false.   □

Problems

5.4. Is there a polyhedron with an odd number of faces, each face with an odd number of sides?

5.5. Prove that the maximum number of bishops that can be placed on an $n \times n$ board without attacking each other is equal to $2n - 2$. (Two bishops attack each other if they are on the same diagonal of the board.)

5.6. Find the minimal number of rooks that have to be placed on a chessboard so that every square is attacked by at least two rooks. (A rook attacks the square it is on and all the squares in the same row and column, including the ones blocked or occupied by other rooks.)

5.7. A cycle of acquaintances is a group of n people such that the first and the second persons are acquainted, the second and the third persons are acquainted, ... , and the nth and the first persons are acquainted.

   Prove that any party of people in which any cycle of acquaintances consists of an even number of people can be partitioned into two groups such that either group consists of mutual non-acquaintances.

5.8. Prove the converse of the statement of Problem 5.7.

5.2 How Can Coloring Solve Mathematical Problems?

5.9. A bureaucratic institution has exactly one entrance and one exit, and a door in the middle of every interior wall of every room (see the floor plan on Figure 5.1.)

   In order to receive a certificate, one must enter the building, visit every room exactly once, and exit the building. Is there a way to receive a certificate?

Solution. At the start, all the rooms seem to be alike. We are going to spoil this equality. Let us color (!) the plan in a chessboard fashion (see Figure 5.2).

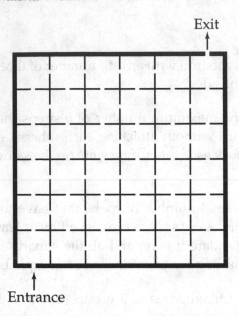

Fig. 5.1. Floor plan of the Institution

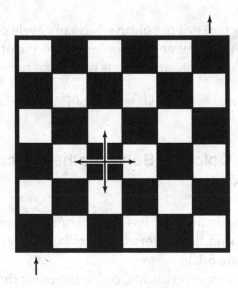

Fig. 5.2.

This coloring has a nice property: from a black room, one can only go into a white room, and from a white room, one can only enter a black room.

While walking through the building we will therefore alternate colors of rooms: black → white → black → white → ....

Assume that there is a required walk passing through every room once, beginning at the entry corner and ending at the exit corner. Since this walk begins at a black room and alternates colors of the rooms on the way, it must end in a white room (we have equal numbers of black and white rooms!), but the exit room is black. We arrive at a contradiction, showing that we cannot get a certificate out of this institution.

What else would you expect from bureaucracy! □

We say that a figure F is *tiled* if it is completely covered by tiles without any tiles overlapping or sticking out of the boundary of F.

5.10. Find all pairs (m, n) of positive integers such that the $m \times n$ checkerboard can be tiled by linear k-ominoes (i.e., rectangles of the size $1 \times k$).[1]

Solution I

If m or n is a multiple of k, then the $m \times n$ board can be tiled by linear k-ominoes (show!).

Assume now that neither m nor n is a multiple of k, but the board can still be tiled by linear k-ominoes. Let us color the board diagonally in k colors with cyclic permutation of colors (see Figures 5.3 and 5.4).

This coloring (*diagonal cyclic k-coloring*) has a remarkable property: no matter how a linear k-omino is placed on the board, horizontally or vertically, it will cover exactly one square of each of the k colors!

Since by assumption the board can be tiled by linear k-ominoes, and every k-omino covers an equal number of squares of every color (i.e., one square of each color), the board contains an equal number of squares of each of the k colors. It is not difficult to prove (do!) that for any positive integers m and k there are nonnegative integers q and r_1 such that $0 < r_1 \leq \frac{k}{2}$ and

$$m = kq + r_1$$

or

$$m = kq - r_1.$$

---

[1] This problem with its three solutions first appeared in the author's article [7].

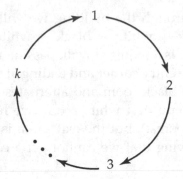

Fig. 5.3. Cyclic permutation of k colors

Fig. 5.4. Diagonal cyclic k-coloring for $k = 3$

Accordingly, we will consider two cases.

1:   $m = kq + r_1$. We cut the given board into two rectangular boards of sizes $kq \times n$ and $r_1 \times n$. Since the $kq \times n$ board can be tiled by linear k-ominoes, it contains an equal number of squares of each color. The given board also contains an equal number of squares of each color; therefore, their difference, which is the $r_1 \times n$ board, has the same property.

2:   $m = kq - r_1$. In this case, we attach the $r_1 \times n$ board to the given board to obtain the $kq \times n$ board and extend the coloring of the given board to a diagonal cyclic k-coloring of the $kq \times n$ board. The $kq \times n$

board can be tiled by linear k-ominoes; therefore, it contains an equal number of squares of each color. The given $m \times n$ board contains an equal number of squares of each color too; therefore the $r_1 \times n$ board contains an equal number of squares of every color.

Thus, in both cases we get the $r_1 \times n$ board with a diagonal cyclic k-coloring, which contains an equal number of squares of each color. Let us rotate this board through a 90° angle; i.e., consider the $n \times r_1$ board, and apply to it all of the above reasoning. As a result, we will get a non-empty $r_2 \times r_1$ board, where $0 < r_2 < \frac{k}{2}$, which contains an equal number of squares of each color.

On the other hand, the number of colored diagonals in the $r_2 \times r_1$ board is equal to $r_2 + r_1 - 1$, and

$$r_2 + r_1 - 1 \le \frac{k}{2} + \frac{k}{2} - 1 < k.$$

Therefore, at least one of the k colors is not present at all in the $r_2 \times r_1$ board!

Thus, we proved that an $m \times n$ board can be tiled by linear k-ominoes if and only if m or n is a multiple of k.   □

The following notation will be helpful to us in the second solution of this problem: *instead of writing "the remainders upon dividing numbers $a_1, a_2, \ldots, a_n$ by n are equal" we will simply write:*

$$a_1 \equiv a_2 \equiv \cdots \equiv a_n \pmod{n}.$$

It is easy to prove (do!) that $a_1 \equiv a_2 \pmod{n}$ if and only if $a_1 - a_2$ is a multiple of n.

Solution II

Assume that neither of m, n is a multiple of k; i.e.,

$$m = kq_1 + r_1, \quad 0 < r_1 < k$$
$$n = kq_2 + r_2, \quad 0 < r_2 < k,$$

but the board is tiled by linear k-ominoes.

Let us color each of the columns of the board with one of the k colors with cyclic permutation of colors (see Figures 5.3 and 5.5), and denote by $S_1, S_2, \ldots, S_k$ numbers of squares of the board colored in the 1st, 2nd, ..., k-th colors, respectively.

This coloring (*column cyclic k-coloring*) possesses a useful property: if a linear k-omino is placed on the board vertically, it covers k squares of the same color; if it is placed on the board horizontally, it covers exactly one square of each color. By assumption, the board is tiled by linear k-ominoes; therefore,

$$S_1 \equiv S_2 \equiv \cdots \equiv S_k \quad (\text{mod } k). \tag{$*$}$$

On the other hand, note that we have one more column of the r_2th

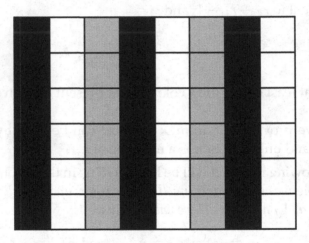

Fig. 5.5. Column cyclic k-coloring for $k = 3$

color than of the $(r_2 + 1)$st color, i.e.,

$$S_{r_2} - S_{r_2+1} = m.$$

Since m is not divisible by k, this proves that S_{r_2} is not congruent to S_{r_2+1} (mod k) in contradiction to the congruence $(*)$ above. This contradiction proves that divisibility of m or n by k is a necessary condition for the $m \times n$ board to be tileable by linear k-ominoes. It is also sufficient. □

We will now discuss the same problem for the third time (see Problem 1.23 and Problem 5.3) and present a solution of it obtained by the winner of the First Annual Colorado Springs Mathematical

Olympiad, Russel Shaffer, during the competition in 1984, and independently by Bob Wood from Colorado Springs and Luc Miller from France in 1986. Bob added symmetry and elegance to the solution by introducing the idea of a cylinder.

5.11. Forty-one rooks are placed on a 10×10 chessboard. Prove that you can choose five that do not attack each other.

Solution III

Let us make a cylinder out of the chessboard by gluing together two opposite sides of the board. Color the cylinder diagonally in 10 colors (see Figure 5.6).

Now we have $41 = 4 \times 10 + 1$ pigeons (rooks) in 10 pigeonholes (one-color diagonals). Therefore, there is at least one hole containing at least 5 pigeons. But the 5 rooks located on the same one-color diagonal do not attack each other!   □

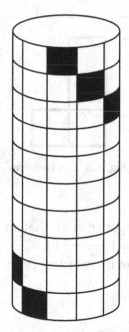

Fig. 5.6. One of the ten one-color diagonals is shown in black.

Problems

5.12. Given an $m \times n$ rectangular chessboard and dominoes (linear k-ominoes with $k = 2$). A pair of distinct squares of the board is called *good* if the figure resulting from the cutting of these two squares out of the board can be tiled by dominoes. Find all the *good* pairs.

5.13. Find all positive integers n such that an $n \times n$ chessboard can be tiled by T-tetrominoes (see Figure 5.7).

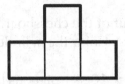

Fig. 5.7. T-tetromino

Will the answer change if you tile a cylindrical board made out of the flat $n \times n$ board?

Fig. 5.8. L-tetromino

5.14. Find all positive integers n such that an $n \times n$ chessboard can be tiled by L-tetrominoes (see Figure 5.8).

Will the answer change if you tile a cylindrical board made out of the flat $n \times n$ board?

5.3 Combinatorics of Sets

We will use standard notation of sets:

$a \in A$ denotes "a is an element of the set A."

$a \notin A$ denotes "a is not an element of the set A."

$A \bigcup B$ denotes the union of sets A and B.

$A \bigcap B$ denotes the intersection of sets A and B.

$A \backslash B$ denotes the set of all elements x such that $x \in A$ and $x \notin B$.

$|A|$ denotes the number of elements of a finite set A.

Let n and k be nonnegative integers with $k \leq n$. The symbol $\binom{n}{k}$ denotes the number of k-element subsets in an n-element set.

5.15. Prove that for any nonnegative integers n and k, $k \leq n$, the following equality holds:

$$\binom{n+1}{k+1} = \binom{n}{k} + \binom{n}{k+1}.$$

Solution. Let I be the number of $(k+1)$-element subsets of an $(n+1)$-element set. Let us fix one element a of the given $(n + 1)$-element set S. We can partition all $(k + 1)$-element subsets into those which contain a and those which do not.

If a is one element of a subset L, which consists of $k + 1$ elements, we need to select k more elements out of $S\backslash\{a\}$, so the number of such subsets L is by definition $\binom{n}{k}$.

If a is not an element of a subset M, which consists of $k + 1$ elements, then we need to select all $k + 1$ elements out of the set $S\backslash\{a\}$, so the number of such subsets M is $T = \binom{n}{k+1}$.

So the total number T of $(k+1)$-element subsets of the given $(n+1)$-element set S is $\binom{n}{k} + \binom{n}{k+1}$.

On the other hand, it is also equal to $\binom{n+1}{k+1}$.   □

5.16. Prove that the following equality holds for any positive integer n:

$$\binom{n}{0} + \binom{n}{1} + \binom{n}{2} + \cdots + \binom{n}{n} = 2^n.$$

Solution. (a) Let S be an n-element set. Since $\binom{n}{0}$ denotes the number of 0-element subsets of S, $\binom{n}{1}$ denotes the number of 1-element subsets of S, $\binom{n}{2}$ denotes the number of 2-element subsets of S, etc., the sum

$$\binom{n}{0} + \binom{n}{1} + \binom{n}{2} + \cdots + \binom{n}{n}$$

is equal to the total number $|P(S)|$ of subsets of the set S.

(b) All that is left to prove is that $|P(S)|$ is equal to 2^n. One way to define a subset S_0 of the set S is to mark each of the n elements of S according to whether or not it is an element of S_0.

Elements	1st	2nd	$\cdots$	nth
Yes or No				

Fig. 5.9.

Thus, in Figure 5.9, we have two options (yes or no) for the 1st block, two options for the second block, ..., two options for the nth block, i.e.,

$$|P(S)| = \overbrace{2 \cdot 2 \cdots \cdots 2}^{n \text{ factors}}. \qquad \square$$

5.17. (*MATHCOUNTS National Competition, 1985*) Given a collection of letters consists of n D's and r C's. Find the number P of different words (sequences) that can be formed from the D's and C's if each sequence must contain n D's (and not necessarily all C's).

Solution. I would like to share with you a solution I found in May 1985 during a meeting of MATHCOUNTS judges (the original solution was quite long). First, let us add one more letter D and find out the number of words that can be composed of *exactly* $n + 1$ letters D and r letters C.

These words have $r + n + 1$ letters:

1st	2nd	$\cdots$	$(r+n+1)$st

In order to uniquely determine a word, we have to pick the positions of letters D and fill the rest with C's, i.e., we need to pick an $(n+1)$-element subset out of the $(r+n+1)$-element set $\{1, 2, \ldots, r+n+1\}$. Therefore, the number R of such words is

$$R = \binom{r+n+1}{n+1}.$$

All that is left to do is notice that the required number P is equal to R.

Indeed, let us take a word W consisting of $n + 1$ letters D and r letters C:

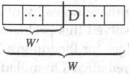

We locate the first D from the right and cut out the end of the word beginning with this D. The remaining part W' of the word W certainly contains n letters D and no more than r letters C. It is easy to show (do!) that two distinct words W_1, W_2 produce distinct remaining parts W_1', W_2', thus completing the proof that

$$P = R = \binom{r + n + 1}{n + 1}.$$

So, adding an extra D and later cutting it off proves to be productive!   □

In 1997, this solution of Problem 5.17 was reproduced by the prolific mathematical author Ross Honsberger as a chapter, "A Brilliant 1–1 Correspondence" in his book *In Polya's Footsteps* [5]. Honsberger writes: "Alexander Soifer has pointed out a brilliant 1–1 correspondence."

Problems

5.18. Prove that the following equality holds for any nonnegative integers n and k, $k + 1 < n$:

$$\binom{n + 2}{k + 2} = \binom{n}{k + 1} + \binom{n}{k + 2}.$$

5.19. How many ways are there to deliver 7 letters if there are 3 couriers and any letter can be given to any courier?

5.20. Given n points on a circle, how many quadrangles (not necessarily convex) can be inscribed in the circle with the vertices at the given points? How many of them are convex?

5.21. Find the number of intersections of diagonals of a convex n-gon P inside P if no two diagonals are parallel and no three intersect in one point. Find the number of intersections of diagonals of P outside of P.

5.4 A Problem of Combinatorial Geometry

I would like to conclude these notes by offering you a problem. In 1970, when I raised and solved this problem, its main part was originally selected by the judges for the 9th-grade competition of the Soviet Union National Mathematical Olympiad, but Professor A.N. Kolmogorov did not approve it, as it was too difficult for a four-hour, five-problem competition. "I am not sure I would have solved it," said the great mathematician. Of course he was right.

> Find all positive integers n such that every triangle can be cut into n similar triangles.

This and other problems of combinational geometry will be discussed in my next book, *How Does One Cut a Triangle?*[2]

The main goal of this book will be to demonstrate synthesis, to show how ideas from various areas of mathematics such as geometry, algebra, trigonometry, linear algebra, and extensions of rings come together in the solution of a problem, and give a mini-model of mathematical research.

---

[2] This book [9] appeared in 1990. A new edition [12] is being published by Springer in 2009.

6

Chess 7 × 7: A Journey from Ramsey Theory to the Olympiad to Finite Projective Planes

New Olympiad problems occur to us in mysterious ways. This problem came to me one summer morning in 2003 as I was reading an unpublished 1980s manuscript of a Ramsey Theory monograph while sitting by a mountain lake in the Bavarian Alps. It all started with my finding a hole in a lemma, which prompted the construction of a counterexample (part (b) of the present problem). Problem (a) is a corrected particular case of that lemma, translated, of course, into the language of a story of a chess tournament. I found three distinct solutions to part (a) and an even more special solution to part (b). As a result, this problem became the most beautiful Olympiad problem I have ever created. It outshined even my lifelong favorite, the 41 Rook Problem, which appears in this book three times: as Problem 1.23, 5.3, and 5.11. Moreover, the journey that led me from Ramsey Theory to problems of mathematical Olympiads continued to finite projective planes!

Chess 7 × 7 (*21st Colorado Mathematical Olympiad, April 16, 2004, A. Soifer*)

(a) Each member of two 7-member chess teams is to play once against each member of the opposing team. Prove that as soon as 22 games have been played, we can choose 4 players and seat them at a round table so that each pair of neighbors has already played.

(b) Prove that 22 is the least possible number of games; i.e., after 21 games the result of (a) cannot be guaranteed.

A. Soifer, *Mathematics as Problem Solving*, DOI: 10.1007/978-0-387-74647-0_6,
© Alexander Soifer 2009

(a) Solution I

This solution exploits an algebraic description of convexity.

Given an array of real numbers $x_1, x_2, \ldots, x_7$ of arithmetic mean $\bar{x}$. A well-known inequality (which can be derived from the arithmetic-geometric mean inequality) states that

$$\sqrt{\frac{\sum_{i=1}^{7} x_i^2}{7}} \geq \bar{x}.$$

Thus,

$$\sum_{i=1}^{7} x_i^2 \geq 7\bar{x}^2.$$

This inequality defines "convexity" of the function $f(x) = x^2$, which easily implies convexity of a binomial function $\binom{x}{2} = \frac{1}{2}x(x-1)$, i.e.,

$$\sum_{i=1}^{7} \binom{x_i}{2} \geq 7\binom{\bar{x}}{x}. \tag{6.1}$$

Observe that above we defined the *binomial function* $\binom{x}{2}$ on all real x (not just positive integers). Also, in a certain informality of notation, we could use $\binom{x}{2}$ not only as a number, but also as a set of all 2-element subsets of the set $\{1, 2, \ldots, x\}$.

Let us name the players in each chess team by integers $1, 2, \ldots, 7$. A game between player i of the first team and player j of the second team can conveniently be denoted by an ordered pair (i, j). Assume that the set G of 22 games has been played.

Denote by $S(j)$ the number of games played by player j of the second team: $S(j) = |\{i : (i, j) \in G\}|$. Obviously, $\sum_{i=1}^{7} S(j) = 22$.

For a pair (i_1, i_2) of the first team players, denote by $C(i_1, i_2)$ the number of second team players who have played with *both* players of this pair: $C(i_1, i_2) = |\{j : (i_1, j) \in G \land (i_2, j) \in G\}|$. The sum of all $C(i_1, i_2)$ together, $T = \sum_{(i_1,i_2)\in\binom{7}{2}} C(i_1, i_2)$, counts the number of triples (i_1, i_2, j) such that each of the first team's players i_1, i_2 has played with the same player j of the second team. This number T can be alternatively calculated as follows: $T = \sum_{j=1}^{7} \binom{S(j)}{2}$. Therefore, we get the equality $\sum_{(i_1,i_2)\in\binom{7}{2}} C(i_1, i_2) = \sum_{j=1}^{7} \binom{S(j)}{2}$. In view of the convexity inequality (6.1), we finally get

$$\sum_{(i_1,i_2)\in\binom{7}{2}} C(i_1, i_2) = \sum_{j=1}^{7} \binom{S(j)}{2}$$

$$\geq 7\binom{\frac{\sum_{j=1}^{7} S(j)}{7}}{2} = 7\binom{\frac{22}{7}}{2} > 7\binom{3}{2} = \binom{7}{2},$$

i.e.,

$$\sum_{(i_1,i_2)\in\binom{7}{2}} C(i_1, i_2) > \binom{7}{2}.$$

We got the sum of $\binom{7}{2}$ nonnegative integers to be greater than $\binom{7}{2}$. Therefore, at least one of the summands is at least 2: $C(i_1, i_2) \geq 2$. In our notation, this means precisely that the pair of first team players i_1, i_2 played with the same two (or more) players j_1, j_2 of the second team. Surely you can seat these four players at a round table in accordance with the problem's requirements!   □

(a) Solution II

This solution harnesses the power of combinatorics.

In the Colorado Mathematical Olympiad's selection and editing process, Dr. Col. Robert Ewell suggested using a 7 × 7 table to record the games played. Armed with his suggestion, I found this solution. We number the players in both teams. For each player of the first team, we allocate a row of the table and for each player of the second team, a column. We place a checker in the table in location (i, j) if player i of the first team has played player j of the second team (Fig. 6.1).

If the required four players are found, this would manifest itself in the table as a rectangle formed by four checkers (a *checkered rectangle*)! The problem thus translates as follows:

A 7 × 7 table with 22 checkers must contain a checkered rectangle.

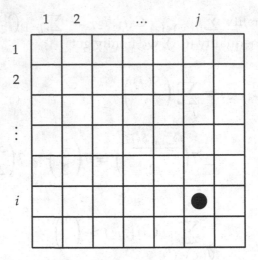

Fig. 6.1.

Assume that a table with 22 checkers does not contain a checkered rectangle. Since 22 checkers are contained in 7 rows, by the Pigeon-hole Principle, there is a row with at least 4 checkers in it. Observe that interchanging rows or columns does not affect the property of the table to have or not to have a checkered rectangle. By interchanging rows, we make the row with at least 4 checkers to be the first row. By interchanging columns, we make all checkers of the first row appear consecutively starting with the first left cell. We consider two cases:

1:   The top row contains exactly 4 checkers (Figure 6.2).

Draw a vertical line L after the first 4 columns. To the left of L, the top row contains 4 checkers, and all other rows contain at most 1 checker each (otherwise, we would have a checkered rectangle that includes 2 checkers from the top row). Therefore, to the left of L we have at most $4 + 6 = 10$ checkers. This leaves at least 12 checkers to the right of L. Thus, at least one of the three columns to the right of L contains at least 4 checkers. By interchanging columns and rows, we put these 4 checkers in the fifth column in positions shown in Figure 6.2. Then the sixth and seventh columns contain at most 1 checker each in rows 2 through 5 (otherwise we would have a checkered rectangle that includes 2 checkers from the fifth column). Thus we have at most $4 + 1 + 1 = 6$ checkers to the right of L in rows 2 through 5. Therefore, in the lower right 2×3 part C of the table, we have

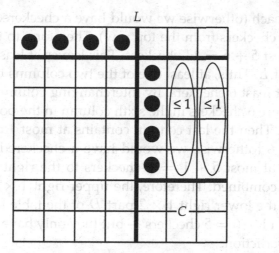

Fig. 6.2.

at least $22 - 10 - 6 = 6$ checkers. Thus C is completely filled with checkers and we get a checkered rectangle in C in contradiction to our assumption.

2:   The top row contains at least 5 checkers (Figure 6.3).

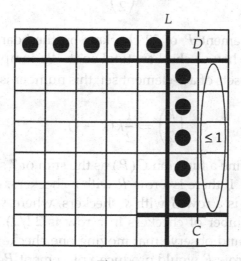

Fig. 6.3.

Draw a vertical line L after the first 5 columns. To the left of L, the top row contains 5 checkers, and all other rows contain at most

1 checker each (otherwise we would have a checkered rectangle that includes 2 checkers from the top row). Therefore, to the left of L we have at most $5 + 6 = 11$ checkers. This leaves at least 11 checkers to the right of L. Thus, at least one of the two columns to the right of L contains at least 6 checkers. By interchanging columns and rows we put 5 of these 6 checkers in the sixth column in the position shown in Figure 6.3. Then the last column contains at most 1 checker in rows 2 through 6 (otherwise we would have a checkered rectangle). We thus have at most $5 + 1 = 6$ checkers to the right of L in rows 2 through 6 combined. Therefore, the upper right 1×2 part C of the table plus the lower right 1×2 part D of the table have together at least $22 - 11 - 6 = 5$ checkers — but they only have 4 cells. We thus get a contradiction.   □

(a) Solution III

This solution is the shortest of the three. It also explains the meaning of the number 22 in the problem:

$$22 = \binom{7}{2} + 1.$$

Given a placement P of 22 checkers on the board, we pick one row; let this row have k checkers total on it. We compute the number of 2-element subsets of a k-element set; this number is denoted by $\binom{k}{2}$ and is

$$\binom{k}{2} = \frac{1}{2}k(k-1).$$

Now we can define a function $C(P)$ as the sum of 7 such summands $\binom{k}{2}$, one per row. If there is a row R with r checkers, where $r = 0, 1,$ or 2, then there is a row S with s checkers, where $s = 4, 5, 6$ or 7 (the average number of checkers in a row is 22/7). We notice that $s - r - 1 \geq 0$, and observe that moving one checker from row S to any open cell of row R would produce a placement P_1 with a reduced value of our function, $C(P) > C(P_1)$, because

$$\binom{r}{2} + \binom{s}{2} - \binom{r+1}{2} - \binom{s-1}{2} = s - r - 1 \geq 0.$$

By moving one checker at a time, we will end up with the final placement P_k, where each row has 3 checkers except one that has 4. For the final placement, $C(P_k)$ can be easily computed: $6\binom{3}{2} + \binom{4}{2} = 24$. Thus, for the original placement P, $C(P)$ is at least 24.

On the other hand, the total number of 2-element subsets in a 7-element set is $\binom{7}{2} = 21$. Since $24 > 21$, there are two identical 2-element subsets (see Figure 6.4) among the 24 counted by the func-

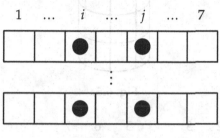

Fig. 6.4.

tion $C(P)$. But the checkers that form these identical pairs form the desired checkered rectangle!   □

(b) Solution

Glue a cylinder(!) out of the 7×7 board, and put 21 checkers on all squares of the 1st, 2nd, and 4th diagonals (Figure 6.5 shows the cylinder with one checkered diagonal; Figure 6.6 shows the cylinder with all three cylinder diagonals, in flat representation).

Assume that 4 checkers form a rectangle on our 7×7 board. Since these 4 checkers lie on 3 diagonals, by the Pigeonhole Principle, 2 checkers lie on the same (checker-covered) diagonal D of the cylinder. But this means that *on the cylinder* our 4 checkers form a square! Two other checkers a and b (which are not on D) thus must be symmetric to each other with respect to D, which implies that the diagonals of the cylinder that contain a and b must be symmetric with respect to D. But no two checker-covered diagonals in our checker placement are symmetric with respect to D. To see this, observe Figure 6.7, which shows the top rim of the cylinder with bold dots for checkered diagonals; distances in squares between the checkered diagonals, clockwise, are 1, 2, and 4. (Distances between the symmetric

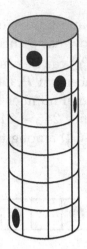

Fig. 6.5.

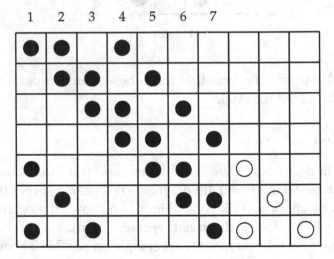

Fig. 6.6.

diagonals and D have to be equal, whereas the numbers 1, 2, and 4 are all distinct.) This contradiction implies that there are no checkered rectangles in our placement. Done!   □

Remark on Problem (b). Obviously, any solution to Problem (b) can be presented in a form of 21 checkers on a 7 × 7 board (left 7 × 7 part with 21 black checkers in Figure 5.9). It is less obvious that the solution is *unique*: by a series of interchanges of rows and columns, any solution to this problem can be brought to precisely the one I presented! Of

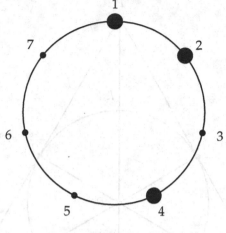

Fig. 6.7.

course, such interchanges mean merely renumbering players of the
same team. The uniqueness of the solution to problem (b) is another
way of stating the uniqueness of the projective plane[1] of order 2, the
so called *Fano Plane*,[2] denoted by $PG(2, 2)$. The Fano Plane is an ab-
stract construction with symmetry between points and lines; it has 7
points and 7 lines (think of rows and columns of our 7 × 7 board as
lines and points, respectively), with 3 points on every line and 3 lines
through every point (Figure 6.8). Observe that if in our 7 × 7 board
(left side of Figure 6.6) we replace checkers by 1 and fill the rest of
the squares by zeroes, we would get the incidence matrix of the Fano
Plane.   ☐

---

[1] A finite projective plane of order n is defined as a set of $n^2 + n + 1$ points with the
properties that:

(1) any two points determine a line,
(2) any two lines determine a point,
(3) every point has $n + 1$ lines through it,
(4) every line contains $n + 1$ points.

[2] Named after Gino Fano (1871–1952), the Italian geometer who pioneered the study
of finite geometries.

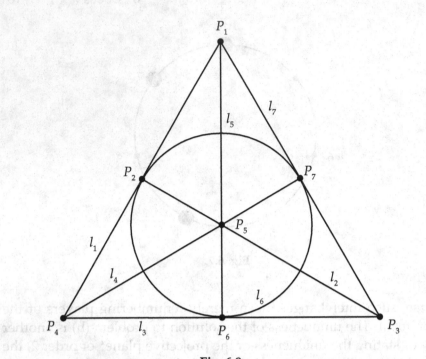

Fig. 6.8.

7

Farewell to the Reader

As Paul Erdős used to say at the end of his lectures, "everything comes to an end," and so has this book. However, if you are inclined to continue your explorations of mathematics with me, I have good news for you. This book is one of seven books that Springer has or soon will publish. If you are interested in a mixture of exciting problems, I recommend you work through the forthcoming book *Colorado Mathematical Olympiad: The First 20 Years and Further Explorations* [14]. The book of "the first 10 years" was published in 1994 [10]. These books also offer a bridge from problems of mathematical Olympiads to problems of "real" mathematics.

If you are receptive to a visual appeal of geometry, you may wish to read the forthcoming expanded edition of *How Does One Cut a Triangle?* [12]. Its first edition [9] was published in 1990. This book offers a glimpse of "real" mathematics, a demonstration of synthesis, where ideas from various branches of mathematics work together to achieve a geometric result.

The forthcoming second edition of *Geometric Etudes in Combinatorial Mathematics* [13] has a dual goal of showing how geometric insight does wonders in service to combinatorics. Its first edition [1] came out in 1991.

Election day, November 4, 2008 (the "yes-we-can" day), saw the release of the book I dreamed of and worked on for 18 years, *The Mathematical Coloring Book: Mathematics of Coloring and the Colorful Life of its Creators* [11]. This voluminous book offers a beautiful mathematics of coloring (so-called *Ramsey Theory*), historical investigations into

A. Soifer, *Mathematics as Problem Solving*, DOI: 10.1007/978-0-387-74647-0_7,
© Alexander Soifer 2009

lives of mathematicians from the Nazi time in Germany to the devastated by World War II Netherlands. The history allowed me to pose questions which have not lost its urgency today, such as the role of a scholar in tyranny. The book presents the aesthetics of mathematics as an art, a philosophy of its foundations, and the psychology of mathematical and historical discovery. The Nobel Laureate Boris Pasternak expressed my goals in this book better and more concisely than I could — great poets often do it well:

> I bring here all: what have I lived through,
> And that what keeps my soul alive,
> My rectitude and aspirations,
> And what have seen my own eyes.[1]

My next book will not include mathematics. However, the great twentieth century mathematician will be the hero of the book, which will be entitled *Life and Fate: In Search of Van der Waerden* [15]. I hope it will be published in 2010.

The book of open problems of the legendary mathematician Paul Erdős will come next, likely in 2011: *Problems of p.g.o.m. Erdős* [4]. I would not have attempted to write it, but in 1990 Paul asked me to help him in the endeavor, and thus it will be our joint book. As you may know, Paul Erdős (1913–1996) was the greatest problem creator of all time. You will be able to work on his problems because very little background knowledge is required to understand many of his problems. Moreover, many problems allow young mathematicians to advance and find partial solutions.

The book after Erdős will be either *The Art on the Frontier of Cultures: The Fang People of West Equatorial Africa and Their Neighbors* or *Memory in Flashback*. The former could be a result of my ongoing study of African art and culture, inspired by the great anthropologist James W. Fernandez. The latter will be a collection of humorous and noteworthy moments of my life.

Having finished this book, you have become my alumnus, the title that carries the responsibility to stay in touch, to send me your most enjoyable solutions, your new problems, suggestions, and ideas. Rest assured, I will always be delighted to hear back from you!

---

[1] Boris Pasternak, "The Waves," 1931, translated by Ilya Hoffman for *The Mathematical Coloring Book*.

References

1. Boltyanski, V.G., and Soifer, A. *Geometric Etudes in Combinatorial Mathematics*, Center for Excellence in Mathematical Education, Colorado Springs, 1991.
2. *Curriculum and Evaluation STANDARDS for School Mathematics*, National Council of Teachers of Mathematics, Reston, Virginia, March 1989.
3. Dynkin, E.B., Molchanov, C.A., Tolpygo, A.L., Rozental, A.K., *Mathematicheskie Zadachi (Mathematical Problems)*, Nauka, Moscow, 1965 (Russian).
4. Erdös, P., and Soifer, A., *Problems of p.g.o.m. Erdős*, Springer, 2011, to appear.
5. Honsberger, R., *In Pólya's Footsteps: Miscellaneous Problems and Essays*, Mathematical Association of America, Washington DC, 1997.
6. Montaigne, Michel de, *Essayes, John Florio's Translation*, The Modern Library, New York.
7. Soifer, A., *Kletchatye Doski i Polimino (Checker Boards and Polyomino)*, Kvant #11 (1972), pp. 2–10 (Russian).
8. Soifer, A. and Slobodnik, S.G., *Problem M236*, Kvant #12 (1973), p. 29 (Russian).
9. Soifer, A., *How Does One Cut a Triangle?* Center for Excellence in Mathematical Education, Colorado Springs, 1990.
10. Soifer, A., *Colorado Mathematical Olympiad: The First Ten Years and Further Explorations*, Center for Excellence in Mathematical Education, Colorado Springs, 1994.
11. Soifer, A., *The Mathematical Coloring Book: Mathematics of Coloring and the Colorful Life of its Creators*, Springer, New York, 2009.
12. Soifer, A., *How Does One Cut a Triangle?* 2nd edition, Springer, New York, 2009, to appear.

13. Soifer, A., *Geometric Etudes in Combinatorial Mathematics*, 2nd edition, Springer, New York, 2009, to appear.
14. Soifer, A., *Colorado Mathematical Olympiad: The First Twenty Years and Further Explorations*, Springer, New York, 2009, to appear.
15. Soifer, A., *Life and Fate: In Search of Van der Waerden*, Springer, 2010, to appear.

Printed in the United States
By Bookmasters